Fractions, Decimals, Ratios, and Percents

Compliments of
NPRIME

Wisconsin Educational Communications Board

A Project supported by the University of Wisconsin System Higher Education Eisenhower Professional Development Program

WRITING COLLABORATORS

Susan Aitken
Richard Bevacqua
Debra Coggins
Leslie Crenna
Gail Dawson
Sue Dirlam
Rosita Fabian
Judy Flores
Sharon Friedman
Joann Fuller
Donna Goldenstein
Rosalyn Haberkern
Amy Hafter
Susan Hennies
Bernadette Homen
Babette Jackson
Bruce Jackson
Kathleen Lloyd
Rose Lock
Jennifer McConnell
Joanne Matala
Eric Moskowitz
Judy Norman
Barbara Polito
Alma Ramirez
Lee Tempkin
David Thornley
Kim Tolley
Hardy Turrentine
Ricki Wortzman
Daniel Zimmerlin

The people listed above contributed to the development of the cases. Although each case has a primary author, all of them were collaboratively framed and revised. We have therefore not cited the author of each case individually. Furthermore, all names in the cases have been changed to protect the identity of the students, teachers, and schools portrayed therein.

FACILITATOR'S DISCUSSION GUIDE

Fractions, Decimals, Ratios, and Percents

Hard to Teach and Hard to Learn?

Edited by

Carne Barnett
Far West Laboratory for Educational Research and Development

Donna Goldenstein
Hayward Unified School District

Babette Jackson
Hayward Unified School District

Heinemann
Portsmouth, NH

Heinemann
A division of Reed Elsevier Inc.
361 Hanover Street
Portsmouth, NH 03801-3912

Offices and agents throughout the world

This publication is funded by Stuart Foundations and based on work supported by the U. S. Department of Education, Office of Educational Research and Improvement, contract number 400-86-0009. Its contents do not necessarily reflect the views or policies of Stuart Foundations, or the Department of Education, nor does the mention of trade names, commercial products, or organizations imply endorsement by our sponsoring agencies.

Library of Congress Cataloging-in-Publication Data

Fractions, decimals, ratios, and percents: hard to teach and hard to learn? / edited by Carne Barnett, Donna Goldenstein, Babette Jackson.
p. cm.—(Mathematics teaching cases)
Includes bibliographical references.
ISBN 0-435-08357-0.—ISBN 0-435-08358-9 (facilitator's discussion guide)
1. Fractions—Study and teaching (Elementary) 2. Ratio and proportion—Study and teaching (Elementary) I. Barnett, Carne. II. Goldenstein, Donna. III. Jackson, Babette. IV. Title: Hard to teach and hard to learn? V. Series.
QA117.F698 1994
372.7'2—dc20 94-19983
CIP

Editor: Toby Gordon
Production: Vicki Kasabian
Text design: Catherine Hawkes
Cover design: Darci Mehall

Printed in the United States of America on acid-free paper
99 98 97 EB 2 3 4 5 6

Contents

Teaching as Questioning

Acknowledgments

A number of teachers have devoted considerable time and creative energy to this project. I admire their courage in writing candidly about their teaching situations, knowing that the cases would be open to critical analysis by others. Nevertheless, they revealed their thoughts, their problems, and their sometimes flawed decisions so that others might learn. The cases and case discussion process evolved through the thoughtful suggestions given by all of the teachers involved.

The ideas, support, and encouragement of Professor Lee Shulman at Stanford University have been instrumental in the development of this casebook and facilitator's discussion guide. He is responsible for bringing both the case technique and pedagogical content knowledge to the forefront of our thinking in teacher education, and thus provides the inspiration for our work. I am grateful for the vision that Lee provides as a member of our advisory board, and I especially appreciate the personal interest he shows by getting to know teachers in the project and using the cases in his own courses.

I am also indebted to the co-editors of this casebook, Donna Goldenstein and Babette Jackson, for reviewing the manuscript and participating in the revision process. They organized and nurtured the collaboration between Hayward Unified School District and Far West Laboratory that made this work possible. More importantly, they knew how to tap the wisdom of teachers to ensure the credibility and practicality of this work.

Others who have guided the case development process include the members of our advisory board, listed in Appendix C. Their thoughtful contributions have been invaluable. In addition, I want to acknowledge my colleagues at Far West Laboratory. Judy Shulman's work provided the first models for teacher-authored cases and inspired me to pursue the same level of quality. Pam Tyson shares the project responsibilities with me, and I greatly value her perspective and in particular, her contribution to the discussion guide. The administrative assistants for this project, Cary Raglin and Leslie Crenna, put their hearts and hard work into preparing the casebook for publication. The publication was also enhanced by the professional work of editors Joan McRobbie and Joy Zimmerman.

Facilitator to Facilitator

The Values of Case-Based Instruction

▪ *Why use cases to facilitate teacher development?*

The idea of using cases as a teacher development tool was brought to the forefront by Lee Shulman in 1986 and has gained considerable suppport in the teacher-education community. It is a promising approach for many reasons. Current leaders in education research, for example, recognize the importance of personal reflection and social interaction in learning. Participants in case discussions tackle problems together, listen to other points of view, and share experiences. What is most exciting about cases, though, is that teachers often identify very strongly with the situations, and the discussions thus become intense and passionate. The intellectual and emotional engagement that results from case discussions sparks new ideas and challenges old beliefs, while simultaneously providing support and encouragement for participants who are making changes in their teaching.

Case discussions prompt participants to frame problems, analyze situations, and argue the benefits and drawbacks of various alternatives. In this way, they play a critical role in expanding and deepening participants' knowledge of mathematics teaching (pedagogical-content knowledge), which in turn helps participants build upon students' thinking in productive ways. Although there are many *general* teaching and learning issues embedded in the cases, they focus heavily on how *mathematics* is taught and learned. As a result, we find that both teachers' confidence and their competence in the subject area are enhanced through case discussions.

▪ *What underlying assumptions and beliefs are central to facilitating case discussions successfully?*

This case-based approach assumes that participants come to the discussions with valuable knowledge and experiences that will contribute to their professional development. One way for teachers to become their own agents for change is to see themselves as partners in the reform movement. They must be the final authority in determining which practices they will adopt as their own.

Maximum commitment from participants is more likely when their participation is voluntary. In our experience, interest in case discussions grows by word of mouth. Veteran members bring in new members and may even form their own offshoot discussion groups. In this way, case discussions can become a grass-roots effort driven by the participants themselves.

A long-term goal is for the teacher participants to become responsible for undertaking and sustaining their own professional development. That is not to say they shouldn't use other resources, but they should be involved in deciding which resources to tap. They should therefore be involved in every level of decision making about the implementation of case discussions in their school, project, or district.

Teachers must learn to trust their own thinking and decision making as they work toward reforming their math programs. If they lack confidence or a deep understanding of the issues, they are more likely to succumb to the pressures and constraints they will undoubtedly face. It is our belief that ultimately case discussions among classroom teachers will be led by the teachers themselves.

▪ *Why is it useful to discuss a collection of cases?*

Sometimes cases are constructed with the intention that each one will be used independently. Although these cases *can* be used individually, they are intended to be used as a collection. They are grouped and sequenced so that key issues will be encountered in different contexts and from different points of view. For example, several cases focus on understanding and teaching the concept of "a part in relation to the whole." This concept is embedded in the context of fraction, decimal, and percent cases. The cases also show how different teachers approach the "part-whole" idea—and how different students respond. These cases are sequenced so that participants are able to refer back to cases previously discussed and make multiple connections among the ideas presented. Maximum benefit is achieved when a complete series of related cases is discussed.

This strategy stems from Spiro, Coulson, Feltovich, and Anderson's (1988) cognitive flexibility theory, in which they propose that learning in complex domains is best accomplished when ideas are reexamined from a variety of vantage points and in new situations. They claim that this approach makes it possible to establish multiple connections among ideas that on the surface may seem dissimilar and thus leads to more flexible thinking and greater transfer of knowledge from one situation to another.

Organizing a Case Discussion Group

▪ *Who are the most likely candidates for case discussions?*

The most evocative discussions arise in groups with great diversity. Diversity ensures the contrasting points of view that are the impetus for evaluating one's own point of view more closely and more critically. We have found that case discussions appeal to all kinds of teachers, those with considerable professional development experience in mathematics and those with little or none, those with years of teaching experience and those just starting out. Diversity across grade levels leads to valuable conversations about what students are like and about the problems in elementary school versus those in middle school.

- ***Are these cases appropriate for use with preservice teachers?***

Case discussions conducted with preservice teachers are effective learning tools. However, since preservice teachers may have limited teaching experience, they often need to rely more on their own experiences as students to address the issues in a case. If some of them do have experience observing, tutoring, or teaching in classrooms, the discussion will be enhanced all the more. It may be useful to introduce preservice teachers to reform issues through their course work or class readings in conjunction with their participation in case discussions. The suggested readings that follow each case can help provide that background.

Introducing Participants to Case Discussions

- ***When launching a new case discussion group, how do I help participants learn what it means to discuss a case? How do I build a supportive and open environment?***

Case discussions may be an unfamiliar form of professional development for many participants. They may be unaccustomed to being resources for one another and may tend to look to you to provide "answers." They may also be uncomfortable speaking openly about their ideas and beliefs and may be afraid of sounding ignorant, less capable, or less skilled than their peers. Even teachers in the same school often don't have the trust and confidence needed to be open in a case discussion.

One way to introduce people to the approach is to invite new participants to a case discussion group that is already established, or to bring veteran participants to your newly established group to "seed" the process. However, if case discussions are new to your area, you may need to start a group from scratch. If so, a videotape of a case discussion can be very helpful in introducing the process.

Another alternative—a one-day working retreat, preferably held away from school—is highly recommended if the financial resources are available to carry it out. You'll need to arrange for a place to meet, teachers will need to be released from their classrooms for the day, and refreshments will need to be provided. A one-day meeting allows ample time to introduce the idea of cases in an informal yet professional manner. It also provides opportunities to nurture the necessary relationships that will help participants be candid and supportive with one another.

The first meeting is critical in setting the tone and building ownership in the process. You will need extra time to provide informal opportunities for participants to get to know one another, take part in a case discussion, and reflect on the process. You may also want to invite veteran case discussion participants to share their experiences. This first meeting should ideally be at least three or four hours long.

■ *How can I help participants acquire a sense of ownership in the case discussion process?*

Since one of the primary goals of case discussions is for participants to assume ownership in the process, they must have an ongoing involvement in planning and shaping the entire program. This includes having a voice in the quality of the facilitation, the quality of the discussions, and the organization of the meetings. Ownership can be addressed at two levels: the administration of the group and its meetings and the procedural process of each case discussion.

One of the best ways to build a sense of ownership is to involve participants in the decision making. Steering committees partially or totally composed of participants can be set up at a school site, at the district level, or within a project. These committees could take responsibility for helping others organize new case discussion groups, recruiting new participants and case discussion leaders, arranging university or district credit, distributing flyers or holding introductory meetings, handling logistics, and so on. (It is, of course, critical that committee members be given the time needed to plan and carry out these responsibilities.)

One way to build ownership within a case discussion group is to solicit oral and written feedback frequently about how the case discussions are going. You can also invite the participants to develop their own ways to monitor the quality of the discussions. (A sample assessment form, developed by the Math Case Methods project, is provided in Appendix A.) By taking participant suggestions seriously and acting on them, they will see that their opinions and ideas are useful and valued. Feeling empowered, participants will assume a more active leadership role in the whole process.

Preparing for a Case Discussion

■ *How much time should a case discussion take? What incentive do participants have for joining a case discussion? When and where should I hold a case discussion?*

Most of these cases can be discussed comfortably in two hours, although the more complex ones may warrant somewhat longer periods. The discussions are commonly held monthly after school and may take place in a participant's classroom, a school library, or a professional development center. If the discussions can be held during the school day, teachers will have more energy to participate in a thoughtful discussion. In our experience, however, this opportunity is seldom available.

Many formats and venues can be successful, and each has its strengths. Discussion groups, for example, may be made up of teachers from the same school or from a number of different schools and districts. They may be composed of teachers at a particular grade level or across a range of grade levels. We find that participants appreciate the cross-fertilization of ideas that occurs when they are from diverse settings, but they also benefit from the increased collegiality that results when discussions are held among teachers at a single school.

▪ *How do I choose which cases to discuss?*

When selecting a case for discussion, consider how strongly the group members identify with reform issues, how much experience they have had with reform issues, how long they have been teaching, and the strength of their mathematics content knowledge. As you become more familiar with the different cases and have had more experience leading discussions of them, the process of selecting cases will become easier.

The set of cases in the first section in the casebook are good "starter" cases. They include a variety of case prototypes and deal with issues that most teachers seem to identify with. Your own judgment and experience, however, are your best guide.

▪ *In what order should I discuss the cases?*

The cases are sequenced so that consecutive cases have at least one common, overlapping issue. The table of contents gives a short synopsis of each case so that participants can easily refer back to cases they have discussed. It also highlights connections to reform documents such as the NCTM *Standards* by referencing key topics from the cases. The cases are *not* sequenced by mathematical topic or grade level.

This method of organization reflects our view on how learning takes place in the complex and messy domain of teaching. As was mentioned earlier, the emergence of similar issues framed in different contexts and viewed from different perspectives increases the probability that learning will transfer from one situation to another.

▪ *What other materials might be useful in case discussions?*

Concrete materials can help participants see mathematics through students' eyes. For example, participants may use materials to illustrate a situation in the case or demonstrate ideas they generate themselves. Therefore, the list of materials that participants referred to during our field tests are included in the discussion guide notes. Some cases include a number of materials, some just one or two, others none at all. As you get to know the members of your discussion group, you will begin to learn which materials are most beneficial to them.

Here is a complete list of the materials cited in this discussion guide:

- Base-10 blocks (ten units by ten units, one unit by ten units, one unit by one unit, and if available, ten units by ten units by ten units)
- Calculators
- Counters (beans or plastic discs, for example)
- Small cubes (one centimeter or one inch)
- Decimal place-value squares (cardboard grids divided into one hundred squares, each with a number of squares shaded to represent a decimal part)
- Fraction pieces, preferably made by the participants themselves by cutting colored paper shapes into halves, thirds, fourths, sixths, eights, sixteenths, and so on
- Grid (or graph) paper in centimeter or half-inch squares, or chart-size grid paper in two-centimeter or one-inch squares
- Pattern blocks (hexagons, trapezoids, parallelograms, and triangles)
- Play money, including one-hundred-, ten-, and one-dollar bills and all commonly used coins
- Plastic cups
- Rulers and tape measures, both U.S. and metric
- String
- Transparency of a grid in centimeter or half-inch squares

■ *What is the purpose of the suggested readings listed with each case?*

The suggested readings that follow each case are not intended to provide solutions to the cases. However, they do contain useful additional information, and can be used as a study resource. Case discussion members may wish to decide for themselves which readings they want to pursue and when. This again reinforces the autonomy and self-direction of the group, and is particularly appropriate for inservice teachers. However, you may decide to assign readings rather than leaving those decisions up to the group.

If you do assign the readings, carefully consider whether to do so before or after the discussion. Assigning a reading before the discussion may foster a more in-depth discussion of the particular issue, but it may also limit the discussion to that perspective. Teachers who agree with the reading will feel validated, while teachers who disagree may either quietly reject or openly resist the ideas. Assigning the reading after the discussion may cause participants to feel either that their contribution to the discussion was inadequate, or that you are trying to convince them of your point of view. Participants might ask themselves, Why spend time discussing the issues, when we will be given the "answers" afterward? Regardless of whether they are assigned before or after the discussion, the suggested readings could encourage participants to refrain from exposing and analyzing their own ideas. On the other hand, the culture and climate of the group can be established in a way that the discussion would not be inhibited.

Many of the readings have been taken from NCTM yearbooks, primarily the 1990 yearbook, *Teaching and Learning Mathematics in the 1990s.* Others are *Arithmetic Teacher* articles. The number of sources have been intentionally limited so that participants will get the most use out of one or two additional readings.

Discussing a Case

■ *What is my role as a facilitator?*

Although you may have more expertise than others in the group, you should not assume the role of the "expert." Instead, your responsibility is to elicit alternative points of view and get participants to analyze these alternatives. You may offer your opinion, but you should probably do so sparingly, since you want the group to develop a strong sense of self-reliance.

In the event that participants accept ideas without considering various perspectives, you may pose alternative views for consideration whether or not you agree with these views. Your goal is not to lead participants to a particular idea or view but to help them honestly evaluate several alternatives and come to their own conclusions about what is best for them and their students. It may be tempting to be impatient with participants who do not share your view of teaching and learning, but it helps to remember that deeply held beliefs do not change overnight. These cases are constructed to offer many opportunities for participants to have their beliefs challenged or reoriented. Each person needs time to reevaluate his or her beliefs and practices in light of what he or she is experiencing in the case discussions and in the classroom.

■ *How could I facilitate with another person?*

If you are a novice facilitator, link up with a partner who can serve as a sounding board when you are planning the discussion and can give you valuable feedback afterward. If, however, you decide to co-lead a discussion, it may be easier if one of you is the primary leader and the other serves as a backup. The backup can step in if the discussion falters or if they notice that someone in the group is not getting an opportunity to participate.

This guide contains notes on each case to help you prepare for a discussion. The issues outlined in these notes are those that were frequently raised in discussions when the cases were pilot tested. The individual case notes are designed to help you strengthen your own understanding of the issues and to anticipate their occurrence in your own case discussion. You may want to pay particular attention to some of the mathematical issues, for example, since the mathematics in these cases can be deceptively complicated. The guide contains space in the margins for you to write personal notes or to record additional issues for future referral.

■ *How and when do I distribute the case to participants?*

Participants will probably prefer to read the case before attending the case discussion meeting. This gives them time to reflect on the case, to mull it over, and to make notes about the issues or ideas involved. As an added benefit, prereading sometimes sparks debate among teachers at a school site before the "official" case discussion meeting. This also gives teachers an opportunity to "try out" the case with their own students before discussing it.

■ *How do I conduct a case discussion?*

There are many ways to conduct a case discussion, depending on your purpose and the members of your group. The format that we developed has four elements: warm-up activities, review of key facts, generation of questions, and the discussion of the case. Each of these elements is discussed in this section.

Warm-ups. Although the primary purpose of a warm-up activity is to "break the ice," it can also serve as a way to introduce the topics that will be discussed. You might, for example, put questions around the room on poster-size paper and ask participants to respond to them in writing. Some questions can be lighthearted, others thought-provoking: How is your body like a fraction? When was the last time you had to divide a fraction? Participants will have fun reading and talking about their responses and become a little more relaxed.

Another way to warm up is to present a starter mathematics problem. Participants new to case discussion tend not to examine the mathematics of the case very closely. They may be insecure about their own mathematics background or may simply relate more strongly to the pedagogical issues. Whatever the reason, it is important to help them become more involved in the mathematics. One way to do this is to pull a mathematics problem from the case and have the participants work on it for a short time before the discussion begins.

Not all cases have a problem that can be excerpted for this purpose, but many do. For example, in the first case, "Take One-Third," the students are asked to draw a picture "where you take one-third of one and one-third." To use this as a starter problem, write it on chart paper and ask case discussion participants to spend a few minutes working on it individually or in pairs. By working on the problem, participants will begin to see the mathematical and pedagogical issues to which it relates. The time allotted for this activity should be enough to draw them into the mathematical issues without undermining the value of the whole-group discussion.

Facts. A comfortable way to begin a discussion is to ask participants to list the facts of the case. This enables them to contribute early in the case discussion, and allows everyone to begin the discussion with the same basic information. It may also be useful to help participants form a consensus about the context in which the case is set. This can be done very quickly, usually in four or five minutes. Spending too much time on it could lessen the overall momentum of the discussion.

Questions raised by the case. After the facts have been listed, you can ask the participants to work in pairs to formulate questions raised by the case. Written issues posed as *questions* are more helpful as a discussion catalyst than *statements.* And questions that refer to specific ideas, examples, or dialogue are usually more productive than more generic questions. Some examples are, What aspects of the teacher's question confused the students? What are the benefits and drawback of using money to teach decimals in this case? The questions are then listed on chart paper for participants to refer to during the discussion. (If participants have already read the case before coming to the meeting, they may have jotted down questions in advance.)

As the pairs of participants formulate their questions, they will probably begin their own minidiscussion. While these minidiscussions are productive, they may turn the larger discussion into an empty "reporting session" if they go on too long. The spirit and spontaneity of the whole-group discussion may be lost. Also, these minidiscussions do not have the benefit of your services as discussion leader, ensuring that assumptions are challenged and alternative perspectives are considered.

Rather than having participants generate their own questions, you may wish to give participants one or two questions that will focus their thinking as they read the case. Examples include, What do you think the student in this case was thinking? What are some of the reasons that this method didn't work?

A word of caution, though: this approach gets the discussion started quickly, but it may inhibit participants from bringing up their own issues. Also, once the discussion has begun, it is often difficult for a participant to redirect the discussion to include an issue that he or she feels is not being addressed. There is a fine line between prompting a discussion and determining its course, thereby undermining the participants' autonomy.

Another risk of focus questions is that unless they are carefully constructed, they can make the discussion feel like an exercise—"Read the paragraph and answer these comprehension questions." Since focus questions are embedded in most of the cases, they often emerge through the voice of a participant, making it unnecessary for the facilitator to take that responsibility.

Discussion path. Most cases in this casebook incorporate compelling problems, and you may be content to let the discussion take its own path, allowing the interrelated issues and ideas to emerge as a natural web.

However, if you prefer a more defined discussion path, you could ask participants to analyze a problem in the case, generate alternative solutions, and then evaluate those solutions. This approach is more effective for cases that have a strong central issue or problem than for those that embed multiple issues that interrelate and conflict with one another.

A few of the cases do not have salient problems. In the beginning, these cases may be confusing to participants since they may not see the point of the case. These cases tend to elicit more analysis than problem solving on the part of participants. For example, the discussion may focus on the consequences of the choices made by the teacher in the case or on the student's thinking, instead of searching for solutions.

Anticipating and Addressing Pitfalls

■ *How can I keep participants from "railroading" an idea or point of view through without considering alternatives?*

It may be helpful to warn participants ahead of time that regardless of what they say, you will push them to defend their ideas and views. Also tell them that you will present alternative views and ask others to do so as well. You can ask such questions as, Why might this be (or not be) a valid point of view? Could someone give an argument in favor of that (or opposed to that)?

It is especially important to bring up alternative perspectives that the participants haven't thought of themselves, that are unpopular, or that are considered inappropriate given current philosophical perspectives in mathematics education communities. Keep in mind that many teachers may hold ideas that conflict with current reform philosophy. Unless these perspectives are carefully analyzed, the discussion may simply reinforce, or will not challenge, what teachers already believe. Discussing these ideas may provoke participants to examine them—and perhaps reject them. If all perspectives are discussed and analyzed in terms of both their benefits and risks, participants will be in a much better position to make informed decisions. In our view, the purpose of case discussions is not to tell participants how to teach, but to help them make informed decisions.

■ *What do I do when participants present a point of view that makes them vulnerable in front of the group?*

It is very important that participants feel that expressing any point of view is encouraged and supported. When someone says something that exposes inexperience, bias, or insensitivity, the rest of the group may become very uncomfortable and the discussion may turn awkward. When this happens, discuss the perspective openly, and invite others in the group to comment. If you prepare the group ahead of time to anticipate this kind of situation, they can learn to be supportive and still disagree. If the other participants do not make supportive comments, you can step in and validate the basis for the unpopular or misguided idea and ask for alternatives. For example, you could say, I understand why you think that way, but is there another way to think about it?

▪ *Why is it important not to stop with the "right answer"?*

It is easy to fall into the trap of listening for what you want to hear, and once you've heard it assume that the mission of the discussion has been accomplished. The danger here is that participants may begin trying to please you rather than exploring their own thinking and reasoning. If you ask participants to justify their responses and to defend all alternatives, they not only may become more articulate about their reasoning, but they may become more convinced of the idea themselves and help to convince others.

▪ *How can I help deepen and focus the discussion?*

One of the most challenging aspects of leading a case discussion is deciding how to push participants to analyze alternative ideas and assumptions in depth. Discussions allowed to wander on their own may disintegrate into superficial chitchat, with ideas being tossed around but not carefully considered or their merit debated. The most valuable discussion is one that focuses on the specifics of the case, not broad generalizations. There are several ways to help draw the group into a deeper discussion. One way is to ask participants to illustrate their ideas on chart paper or with concrete materials. For example, if a participant suggests that using manipulatives will help students understand a particular problem, you could ask, *What* manipulatives? *How* might they be used? Could you show us? Sample questions such as these are provided in the case notes, but they should only be used as occasional prompts. You will find that once the discussion begins, it takes on a life of its own; the group spontaneously generates its own questions.

Listed below are some additional, more generic questions you can keep in mind during the course of the discussion:

Building on Other's Ideas

Does anyone have a comment about that idea?
How does someone else feel about that?
Can someone offer a different way of thinking about that?
Could you think of a reason why someone might have a different point of view?

Justifying Ideas

What in the case makes you think that is so?
What are the assumptions behind that statement?
What justification can be given for that suggestion?

Examining Alternative Points of View

What are the long-term/short-term consequences of this idea?
What are the limitations or advantages of this approach?
What trade-offs do you make by doing this?
How important is this idea or topic?

Analyzing Ideas and Philosophical Underpinnings

What about the teaching might be influencing the way the student is thinking?
What beliefs about mathematics are being communicated by this teacher?
What values are communicated by this form of teaching?
Are the students in this case active or passive learners?
What do we know about the learning community from clues in this case?
What is the teacher doing that promotes or impedes student thinking?
How does this case help or hinder independent thinking by students?

▪ *How do I prevent two or three people from dominating the discussion and encourage less assertive people to join the discussion?*

If some members dominate the discussion—even if it is out of enthusiasm—you may suggest that members of the group raise their hands or nod when they wish to respond. Some participants may need more time to think before joining a discussion and may feel they don't get this opportunity when other members jump in too quickly. Another way to slow the discussion down is to say, "Before you answer, give yourself a minute to think about your response."

Some participants sit quietly during a discussion, preferring to listen rather than speak. This preference may be the result of cultural and/or personality differences and should be respected. Some members may speak very little in the beginning and become very vocal later. For some, it takes longer to adapt to this kind of professional development and to feel ready to risk asking questions or sharing opinions.

▪ *What do I do if there is a lull in the discussion?*

Silence is something that new facilitators always worry about, but it's rarely a problem. Getting the discussion started may take a little time, but once started it seldom slows down. If the discussion seems to be at a turning point but no one is taking it in a new direction, you can suggest a new issue from among those generated at the beginning of the discussion or ask the participants which one they'd like to pursue next. If you're a first-time facilitator, you may want to ask a colleague to serve as your backup, helping revitalize or redirect the discussion if necessary.

■ *What do I do when important issues have not been brought up?*

This is a judgment call. You may decide to let the issue emerge in another discussion, since the cases have been designed with overlapping issues. Or you may decide to pose the question yourself. Be aware, though, that very specific facilitator-prompted questions can be awkward. Since they have not emerged naturally from the discussion, they are interpreted differently. For example, it may seem that you are trying to elicit a particular "answer." Also, if questions are not perceived as honest, they may inhibit genuine discussion; the session then turns into a let's-figure-out-what-the-facilitator-wants-and-give-it-back game.

■ *What do I do when important issues are brought up and we're almost out of time?*

Discussions are sometimes still going strong when time is up. You may want to use an issue that has just been brought up as the starting point for the following meeting. However, by their nature, case discussions often seem "left up in the air," and this unsettled feeling is actually beneficial. Participants are prompted to continue the discussion with their colleagues and to experiment in their classroom. As much learning can occur outside a case discussion as during one.

Closing and Assessing the Discussion

■ *How do I give participants opportunities to bring their own experiences to the discussion?*

Teachers in discussion groups almost always spontaneously bring up situations that have occurred in their own classroom. They'll share lesson materials, examples of student work, and anecdotes, and are very eager to talk about their attempts to try out new ideas or test new theories that relate to the cases. It's very effective to let this type of sharing arise informally as a natural part of the discussion.

However, you may want to set aside a specific time for the participants to share their experiences. You might suggest journal writing activities that encourage them to reflect on their classroom experiences. However, if you do think it's important to take a structured approach to reflection, consider asking participants to help decide what kind of reflection process will be most useful to them.

▪ *How can I help bring closure to the case discussion in a way that helps participants put the ideas and issues into perspective yet doesn't discourage further reflection?*

Perhaps your biggest challenge as a case discussion leader is to help participants reflect on and consolidate ideas that have been sparked during the discussion. Bringing closure does not mean presenting hard-and-fast solutions; rather, it means getting the participants to stand back from the particular discussion and view it from a larger perspective—a perspective that may differ for each participant. The closure process should include articulating and clarifying the issues and ideas raised during the discussion. Here are some ways to go about it (and remember: don't always use the same ones—participants will appreciate the variety).

Reflections in pairs. At the close of the discussions, introduce questions like, What thoughts (ideas, statements, insights) did this discussion trigger? What questions did this discussion leave unresolved? Ask pairs of participants to discuss the question and then report back to the whole group.

Summary. A summary can be done either by you or by a participant who has been taking part in case discussions for some time and feels comfortable with this assignment. The person who summarizes can describe some of the major issues discussed and put them into a broader perspective, reinforcing or consolidating some of the ideas discussed. Also, that person can reframe unresolved questions in a way that invokes further thought and reassures participants that these issues will not be dropped. Often just restating the points of the discussion provokes major shifts in thinking. The comments in this discussion guide should help you summarize the discussion effectively.

Appreciation. Ask the participants to share something they appreciated about the discussion or another participant's contribution to it.

Round-the-table comments. Ask participants to respond briefly in turn to a questions like, What can we learn from this case? What was positive, negative, or interesting about this discussion?

Journals. Near the end of the session, participants may be given time to write about the reflections, insights, concerns, or questions they have. (Some participants may want to do this later, after they've had more time for their ideas to incubate.) The writing can be totally unstructured, or it can be structured around a question of your choosing. Some possible focusing questions include: What ideas or practices have these case discussion confirmed or validated for you? What ideas or practices has this case discussion led you to question or see differently?

How can the discussion process be assessed?

It is valuable to develop a formal or informal way to gather information about the group dynamics and interactions from the participants' point of view. This will ensure that the participants get feedback about how others in the group are feeling about the group interaction and take responsibility for improving the process themselves. The "Case Discussion Process Assessment" form (Appendix A) is an example of a form one facilitator and the members of a discussion group co-developed for this purpose. If you use a form, ask the members of your group to fill it out (in confidence) after the case discussion. You (or some other member of the group) can quickly scan the assessments and share highlights and pertinent comments with the group.

How will I receive feedback?

Whether you are a novice or a very experienced facilitator, you need to get feedback about your methods from the discussion participants. One way is to ask a participant (or a colleague whom you've asked to observe the discussion) to share with the group his or her perception of how things went. Another way is to videotape the discussion and view the tape later with one of the participants or with a colleague. A third way is to give the participants an opportunity to write about their perception of your strengths and what you might do to improve. (You can use the "Facilitator Feedback" form in Appendix B to structure the feedback.)

MATHEMATICS TEACHING CASES

FACILITATOR'S DISCUSSION GUIDE

Fractions, Decimals, Ratios, and Percents

Inside Student Thoughts

Take One-Third

This case is a vehicle for discussing broad philosophical issues, such as beliefs about learning and handling incorrect solutions. It also addresses important pedagogical content issues having to do with the representations of parts and wholes and the concepts underlying multiplication with fractions.

Suggested Materials: Counters, pattern blocks, fraction pieces. (Refer to the introduction of this book for more information about the suggested materials.)

Questions and Issues Raised by the Case

▪ *Language issues*

Students may have been confused by the language the teacher used to present the problem. They may have thought that *take* meant *take away* and just crossed out the $\frac{1}{3}$ to take it away. Should the teacher have avoided using *take* and stated the problem another way? How might this have been done? Do students need to understand the language used in this way at some point?

Students may have also been confused by the word *of*. How do we help them learn that this word means multiplication? Could it be related to how we use the word with whole numbers? For example, "2 sets of 3" could be related to "$\frac{1}{2}$ set of 3." Teachers must remember that the subtleties, precision, and duplication in mathematical language is especially difficult for second-language learners—and the word *of* is very abstract.

▪ *What is the whole?*

In this problem, students had to consider the $1\frac{1}{3}$ to be the whole. Then they had to figure out what $\frac{1}{3}$ of this whole was. This can be confusing since students are accustomed to thinking that the complete circle is the whole. The idea that the whole may be a part of something or a set of things must be developed if students are going to acquire a deep understanding of basic fraction ideas.

▪ *Multiplying by a fraction*

This case offers an opportunity for teachers to examine what it means to multiply by a fraction. They may enjoy the challenge of trying to represent other fraction multiplication problems. How would they represent $1\frac{1}{2}$ groups of $1\frac{1}{3}$, or $\frac{2}{3}$ of $\frac{3}{4}$? Could they make up a word problem to fit these multiplication examples?

Amanda pointed out the relationship between multiplying by a fraction and dividing by a whole number. Why does this work?

■ *Beliefs about teaching*

Some teachers will ask why one would start a lesson on multiplying fractions with a problem that the students had never seen before. Why didn't this teacher begin with some examples first or ask them to review the previous day's work? Wouldn't the students have been more successful if they were shown how to do it first?

This points out the major difference between the conventional "tell-them-how-to-do-it-and-then-have-them-practice" approach and the constructivist approach. Teachers must consider the consequences of both approaches. How might students who are accustomed to either approach differ in their beliefs about mathematics? If students are asked to figure things out for themselves and learn by talking and working with others, how does that impact the way they think about mathematics?

Should the teacher have given a little more guidance? How much guidance can a teacher give and still be teaching from a constructivist point of view? Are there some things that teachers should tell students and other things that they should figure out on their own?

■ *Conceal or reveal students' incorrect solutions?*

The teacher in this case purposely called on several students to present their solutions without first checking to see if the students had the correct solution. Why would the teacher do this? Did she realize that students might be embarrassed? What does it do to a student's self-esteem to display an incorrect solution? Isn't there a danger that many students will remember the wrong solution if it is presented?

What needs to be considered is what students can learn from each other, both from their correct approaches and from their mistakes. However, the learning environment must be conducive to this approach or student self-esteem can be badly damaged. Some students are also very sensitive and must be encouraged to take risks, whether or not they are certain of their solutions. Risk taking and empowerment are interrelated. When students accept a challenge and take a risk, they learn to trust their own capabilities.

Suggested Reading

Borasi, R. 1990. "The Invisible Hand Operating in Mathematics Instruction: Students' Conceptions and Expectations." In *Teaching and Learning Mathematics in the 1990s*, edited by T. J. Cooney and C. R. Hirsch, 174–182. Reston, VA: The National Council of Teachers of Mathematics.

Miller, L. D. 1993. "Making the Connection with Language." *Arithmetic Teacher* 40(6): 311–316.

Beans, Rulers, and Algorithms

This case reveals the complications many teachers experience when they design a unit to help students understand fractions. Simply using manipulatives is not enough to ensure that students are learning. Choosing what representations, experiences, and examples to use and deciding how to present and sequence these experiences is critical to success.

Suggested Materials: Counters (i.e., beans), rulers, fraction pieces.

Questions and Issues Raised by the Case

- ***Confusion of part and whole***

The picture that the teacher drew on the board could be confusing. By drawing circles around the sets of beans, students might have gotten the impression that each of these sets represented a whole.

- ***Discrete or continuous representation?***

What are the advantages—or disadvantages—of beginning with beans (a discrete representation) as opposed to a ruler or a pizza (continuous representations)? The teacher may have chosen to use beans so she could illustrate the common denominator using a whole that was already partitioned into 12 parts. Do students see the 12 beans as a whole? Is it possible for them to find the answer to this problem by thinking in terms of whole numbers? For example, could they simply pay attention to the numerators and add 4 and 3 to get 7 beans? Do they understand that the 4 and 3 represent parts of a whole?

What if a pizza were used for the first examples? A circle has a visual "wholeness." Would this possibly make part-whole relationships easier to keep in mind?

What if a ruler were used for the introduction? Keep in mind that rulers are prepartitioned, and they indicate equivalent fractions. In other words, one can see that $\frac{1}{2}$ is equivalent to $\frac{8}{16}$. Do these characteristics help or hinder the development of the ideas presented in this lesson?

Creating a need to find a common denominator

By designating the whole as 12 beans, the teacher removed the need to find a common denominator. In other words, students didn't experience the dilemma of trying to name the fraction that represented the sum of $\frac{1}{3}$ and $\frac{1}{4}$ and then coming up with twelfths (or twenty-fourths!) for themselves. The twelfths were already given. Will students know why or when a common denominator is needed?

Why did the teacher choose the example, $\frac{1}{3} + \frac{3}{12}$, instead of $\frac{1}{3} + \frac{1}{4}$? Was this an attempt to make the common denominator easy to find? What were the implications of this decision?

Big jumps

It looks like the teacher made a big leap from finding common denominators with concrete materials to problems involving adding fractions that students could not solve using their intuition or common materials. Who adds sixths and sevenths in a real life situation? Are students developmentally ready to add fractions in the fifth and sixth grades? This question is probably too simplistic. Most students probably already understand the basic idea.

Transfer from concrete to abstract

How does one make connections from concrete to abstract? Is it possible that the beans were just as abstract to the students as the pencil and paper algorithm? How could adding fractions be made more concrete? One idea offered during teacher discussion has been to present a problem based on a real-life situation for students to solve using concrete materials or by drawing pictures. Here, students could capitalize on their informal understandings. If the problems were meaningful to them, they could come up with their own ways of representing addition of fractions.

Another suggestion is to ask students to estimate the answer to these problems. For example, students might have to decide if $\frac{1}{4}$ of a pie added to $\frac{1}{2}$ of a pie would be more or less than a whole pie.

Suggested Reading

Yackel, E., P. Cobb, T. Wood, G. Wheatley, and G. Merkel. 1990. "The Importance of Social Interaction in Children's Construction of Mathematical Knowledge." In *Teaching and Learning Mathematics in the 1990s*, edited by T. J. Cooney and C. R. Hirsch, 12–21. Reston, VA: The National Council of Teachers of Mathematics.

There's No One-Half Here

As adults, we may feel secure that we know how to multiply fractions because we know an algorithm for finding an answer to a given problem. However, we may find we are on shaky ground if asked to think of a situation that calls for multiplying fractions or if asked to show we understand that algorithm. This case gives participants opportunities to confront their own misunderstandings and to help each other work though the conceptual underpinnings of fraction multiplication.

Suggested Materials: Counters (i.e., cube blocks).

Questions and Issues Raised by the Case

▪ *Continuous versus discrete representations*

Would the same difficulties have arisen in this case if the teacher had chosen another manipulative for the lesson? For example, what if the model had been a pizza instead of a set of objects such as the cubes? Participants may recognize the differences between these two kinds of models and have informal ways of referring to them. For example, they may refer to the pizza as a whole and the cubes as a set. The more formal vocabulary for a representation such as pizza is a *continuous model* and for representations of sets of things is a *discrete model.* What would be some other examples of continuous and discrete models? How are these models different? Is one model more difficult to use than the other? Should one model be introduced first, or should students be allowed to choose their models?

When using the cubes for the model, the problem could be interpreted as $\frac{1}{2} \times 8 \times \frac{3}{4}$. In other words, the students would be asked to find $\frac{1}{2}$ of 8 and then to find $\frac{3}{4}$ of 4. This is almost like a two step problem, and it could be confusing to some students who may not see the relationship to the original problem, $\frac{1}{2} \times \frac{3}{4}$. Working with a continuous model would not pose the same difficulty.

▪ *Start with a simpler problem*

One suggestion often made to improve this lesson is simplifying the problem presented to the students. If, for example, the students were asked to find $\frac{1}{2}$ of $\frac{1}{2}$, or $\frac{1}{2}$ of $\frac{1}{4}$, they might be able to solve the problem intuitively by drawing or using materials. They might not need to be led through the process step by step as the teacher did in this lesson. Perhaps then the students could be asked to build on their intuitive knowledge to solve more difficult problems, such as the one presented in this lesson, and go on to figure out how to solve other problems involving mixed numbers, such as $\frac{1}{4}$ of $1\frac{1}{2}$.

▪ *Word-problem representation*

Perhaps students in the lesson needed a practical word problem to help them interpret the meaning of multiplying fractions. This suggestion may seem simple on the surface, but many adults struggle when asked to come up with a word problem. Consider this problem, suggested during a discussion:

> *A train has 8 cars, and $\frac{1}{2}$ of them are nonsmoking. Three-fourths of the nonsmoking cars are luxury cars. How many cars are in the luxury section?*

The difficulty is that the answer to this problem (3) is not the same as the answer to the problem in the case ($\frac{3}{8}$). The question would have to be changed to ask, "What part of the whole train are luxury cars?" Most adults will probably struggle to find a word problem that fits the problem in the case and is not relatively contrived.

▪ *Developmental and curricular issues*

Given the difficulty in finding a word problem as mentioned above, a question can clearly be raised about the relative importance of this mathematical topic in the curriculum of a fifth-grade student. How much time should be spent teaching students to multiply fractions? The algorithm is easy, and perhaps students should just be taught how to do the procedure without attempting to understand the underlying concepts. What are the implications of this decision? Can students understand simple multiplication of fraction problems using drawings or concrete materials? Do they need to understand more complicated problems?

For some participants, the argument is not *whether* to teach multiplication of fractions but *when* to teach it. Are these concepts developmentally inappropriate for fifth-grade students? There is speculation that some students this age may have difficulty maintaining a mental representation of the whole while operating on the parts. In other words, the students may not hold onto the idea that the 8 blocks is the whole, while they are figuring out what $\frac{1}{2}$ of $\frac{3}{4}$ of that whole is.

▪ *Language usage*

Was the language that Mrs. Tinley used clear? Participants will often wrestle with clarifying the teacher's language, thinking that this will help students understand. For example, someone might suggest that she try saying: "Take $\frac{1}{2}$ of 8 cubes. How many are left? Now take $\frac{3}{4}$ of that $\frac{1}{2}$." The problem is that this language might also be confused, since *take* and *left* are often associated with subtraction. One alternative that might be more acceptable is: "Show me $\frac{1}{2}$ of your 8 cubes. Now show me $\frac{3}{4}$ of that $\frac{1}{2}$."

There may also be concern about the interchangeable use of the words *times* and *multiplication*. What strategies can be used to help students internalize the different terms? Also, for many students, the use of the word *of* as a multiplication term may seem unfamiliar. One way to help students see the relationship between *of* and *multiplication* is to relate the expression $\frac{1}{2}$ of $\frac{3}{4}$ to an expression that might seem more familiar, such as $\frac{1}{2}$ of 4, or 2 sets of 4. Teachers of second-language learners may be good resources for others in the group who are trying to help students make sense of these terms.

▪ *Meaning of concept and algorithm*

What is a concept or an algorithm? It may be helpful for participants to spend some time discussing these terms, if they arise in the discussion. Some participants, for example, may believe that a concept is equated with a manipulative and that if you are teaching with manipulatives you are teaching concepts. Others may think of concepts as definitions or procedures. For example, $\frac{1}{4}$ may be thought of in terms of the definitions for the numerator and the denominator, rather than a relationship between the amounts represented by the numerator and the denominator.

In this case, the confusion can be even more subtle. For example, was Mrs. Tinley teaching her students: an algorithm to multiply fractions, how to understand the algorithm, or the conceptual underpinnings of fraction multiplication? An algorithm is simply a procedure or a set of rules. In this lesson, Mrs. Tinley did not seem to have this as her goal. She also did not seem to be trying to teach her students how the algorithm worked. Instead, she seemed to be asking her students to explore the conceptual underpinnings of fraction multiplication. She wanted students to know what it meant to multiply fractions, rather than what the meaning of a rule for multiplying fractions was.

Suggested Reading

Witherspoon, M. 1993. "Fractions: In Search of Meaning." *Arithmetic Teacher* 40(8): 482–485.

Everything I Know About Decimals

This case is designed to be discussed in two parts. Begin by having participants read and discuss Part One, which describes a challenge in the teaching of decimals. Then read and discuss Part Two and the Epilogue, which describe what the teacher actually did to address the challenge and the teacher's reflections about her students. Dividing the discussion of this case facilitates the deepening of the discussion. When all are considered at once, there is a danger that the discussion will remain at a general level.

The teacher and the sixth-grade students were working with decimals and place value. Students were using the manipulative of a whole square predivided into smaller squares to represent tenths, hundredths, and thousandths. The teacher wondered whether students were understanding the mathematical concepts or whether they were merely using manipulatives in a rote fashion.

Suggested Materials: Metric tape measures or rulers, base-10 blocks, grid paper, decimal place-value squares.

Questions and Issues Raised by the Case

Part One

- ***Equity issues***

The teacher believed that "children from Latin America who had been to school there usually excelled in math, yet most of our native-born Chicano and African American students were, year after year, failing dismally in this area." Does this imply that Latin-American schooling is better, at least for these students, than is schooling in the United States? Is this difference in school performance a result of important cultural differences that impact student learning? Is it more likely that students born in other countries come from different socioeconomic backgrounds? Foreign-born students who have been to school and who have come to the United States may have more advantages than ethnic minorities born in the United States.

▪ *Rote manipulatives?*

Is it possible to use manipulatives in a way that leads to rote learning? It seems that this was the teacher's concern after reading Ana's journal. Even though Ana had experienced decimals with manipulatives, the writing in her journal could indicate that she was thinking about decimals in a rote manner. How and when do manipulatives reinforce rote learning? Is there anything wrong with rote learning?

Was this manipulative appropriate for the understanding desired? The author of the case placed considerable emphasis on the issue that the squares were predivided. Why did the teacher think that this term was so important? Would it have made a difference if the students were asked to make the divisions themselves? The teacher may have thought that if the students made the divisions themselves, they would be more likely to think of the divisions as parts of a whole. However, when the divisions are already present on the concrete material, students merely have to count the divisions, or they may treat the divisions as whole numbers rather than parts of a whole number.

What are some other materials that are not predivided? What are some strategies that encourage students to make their own divisions?

Is it practical to expect students to divide materials? Is there enough time to divide something into hundredths, for example? Will the task be so tedious that students lose sight of what they are doing?

▪ *What is a decimal number?*

The teacher in this case wanted students to understand that a decimal was part of a whole that had divisions which were multiples of 10. Does a decimal always represent a number of parts out of 10, out of 100, and so on? Could it also represent 2 parts out of 5? On the surface decimals seem fairly simple—they are just fractions with denominators that are multiples of 10. However, upon closer examination, we realize that a decimal is a ratio that describes a comparison of two whole numbers, such as the ratio of 5 compared to 10 (e.g., 0.5), or 4 compared to 20 (e.g., 0.2). Decimals can also be used to describe a quotient of two numbers (e.g., 6 ÷ 24 or 0.25; 5 ÷ 2 or 2.5). What strategies could be used to help students understand that decimals are relationships?

▪ *Was the journal question a good strategy for assessing understanding?*

Was the first journal question too open ended? Isn't it possible that the students knew far more than they could communicate but were unable to do so effectively, due either to obstacles with language or writing, or with the vagueness of the question? If the teacher had asked a more focused question, would the student responses demonstrate more depth in understanding? What was the purpose of this assessment? Was it for the teacher to assign a grade or to begin to get feedback on student knowledge of decimals? Should the teacher grade this assignment?

Based on Ana's journal writing, did the teacher have adequate evidence to be concerned about Ana's conceptualization of decimals? Should the journal assignment be viewed as merely the first step in an assessment process? What other strategies could complement this assessment?

▪ *Ana's view of mathematics*

Ana's journal indicated that she knew some rules about how to work with decimals. She wrote, for example, that you line up the decimals so you get the right answer and that you write a point after a whole number so you know where to "put it" when you subtract. Why might the teacher have been disappointed with Ana's responses? What might this journal entry say about Ana's view of learning mathematics?

What else could the teacher have learned from the journal entries? Did Ana not understand or did she not have the fluency in English to interpret and demonstrate this understanding?

Part Two

▪ *Using metric to teach decimals*

What advantages did the teacher see for using the metric system? Are there also disadvantages? Instead of using metric measuring instruments that were already divided into multiples of 10, the teacher chose to have students make their own instruments. What was the reason for this decision? Would the same things have been accomplished if students had drawn and shaded their own squares in the grid in Part One of this case?

▪ *Use of a single model*

The teacher focused her whole unit on the metric system. Are there advantages to using a single model? What are the dangers? Notice that the teacher hoped to accomplish two things with the unit: She wanted her students to learn about decimals, and she wanted them to learn about the metric system in order to coordinate with two science units.

▪ *Relative size or magnitude of decimals*

The teacher noted that what was measured was not predivided. Why might the teacher have felt this was important in terms of students' understanding of decimals? Does this have anything to do with developing number sense or relative magnitude? Notice that many of the student experiences involved estimating sizes and distances. What was the purpose of these experiences? How were these experiences different from the "shaded squares" experiences in Part One of the case?

Epilogue

■ *Drill and practice versus critical thinking skills*

The teacher questioned whether students needed additional drill on computational aspects of fractions, decimals, and division to do well on their achievement tests. Was the teacher doing a disservice by not spending more time helping students develop facility in rote procedures that they could call upon in testing situations? Are some students better served if they first learn the rote symbol manipulation, and perhaps uncover the underlying conceptual meanings at a later time? Are there individual differences among students that make it easier for some to learn through drill and practice and others through critical-thinking skills?

Suggested Reading

Cuevas, G. 1990. "Increasing the Achievement and Participation of Language Minority Students in Mathematics Education." In *Teaching and Learning Mathematics in the 1990s,* edited by T. J. Cooney and C. R. Hirsch, 159–165. Reston, VA: The National Council of Teachers of Mathematics.

Holmes, E. E. 1990. "Motivation: An Essential Component of Mathematics Instruction." In *Teaching and Learning Mathematics in the 1990s,* edited by T. J. Cooney and C. R. Hirsch, 101–107. Reston, VA: The National Council of Teachers of Mathematics.

Secada, W. G. 1990. "The Challenges of a Changing World for Mathematics Education." In *Teaching and Learning Mathematics in the 1990s,* edited by T. J. Cooney and C. R. Hirsch, 135–143. Reston, VA: The National Council of Teachers of Mathematics.

Steen, L. A. 1990. "Mathematics for All Americans." In *Teaching and Learning Mathematics in the 1990s,* edited by T. J. Cooney and C. R. Hirsch, 130–134. Reston, VA: The National Council of Teachers of Mathematics.

How Can 100% of Something Be Just One Thing?

This case is an excellent vehicle for helping teachers understand how confusing percentage concepts can be from the students' point of view. What seems obvious—that 100% is 1—is not at all obvious to students. Although this case focuses on the preparation of students for a mathematics contest, it should not be limited to this particular context. One way to expand the discussion is to ask teachers to consider how the information in the case might influence the way they would teach a unit on percent in the future.

Suggested Materials: Play money, base-10 blocks.

Questions and Issues Raised by the Case

- *"Trick problems"*

Some teachers may argue that this particular multiple-choice problem was a trick problem and should not be used for any purpose. Other teachers may see this as a good discussion starter for a unit on percent, but they may be opposed to using it for assessment, evaluation, or competition. Others may argue that it is a good question for a contest but should not be used to assess or evaluate student understanding.

- ***Issues of equity and access***

Should this problem be assigned to all students or only those with strong self-esteem? Teachers may argue that assigning this problem to students could cause them to become unsettled, and therefore it should not be used with students who do not see themselves as strong mathematical problem solvers. Others may argue that shielding students from these types of problems is not fair because it limits their access to problems that they may encounter later in life. Since such questions are common in contests, achievement tests, and college entrance exams, should we prepare students to develop a critical eye for them? How might this be done without compromising what we believe are good teaching practices? Teachers may argue that the issue of access is particularly salient for low-income, ethnically diverse populations.

▪ *Part of a unit versus part of a set*

Could we say that 100% of a group of 15 students is 15? Could we say that 100% of a group of 15 students is 1? This is at the heart of the confusion. When you talk about 100% of something you may be talking about 100% of a set (such as 15 students) or of a continuous unit (such as 1 cake). If you think of the 15 students as 1 whole group of students, 100% is 1 group. One strategy for addressing this confusion is to find examples of groups that have names, so that a connection between the part of the set and the part of the unit can be made. For example, one could say 100% of a dozen eggs is 1 dozen or 12 eggs.

Some curriculum materials reinforce the idea that 100% is 100. Often student materials introduce percent concepts with a 10-by-10 square of graph paper. Students are asked to shade in 45% or 10% of the square. Since the square is composed of 100 smaller squares, the task is simple. In this case, 100% of the square is 100. These materials seldom ask students to shade in 10% of a unit that has no subdivisions or 10% of a unit that is divided into 5 equal parts.

What implications does this have for planning a unit on percent? What materials and methods could be used to build these ideas?

▪ *Connection to the real world*

What does 100% mean? Is it meaningless without saying 100% *of* something? After all, 100% of 10 is 10, and 100% of 100 is 100. Should assessments and lessons always include contextual information or connections to students' experiences? Why or why not? How might this be done? Consider asking students to use their intuitive knowledge of percent to draw pictures representing approximately 100%, 50%, 25%, or 10% of something like a jug of water.

▪ *Gap in knowledge or failure to use knowledge?*

Why did the students who knew how to convert 100% to a fraction fail to access that information? Was there really a gap in their understanding? One possible explanation is that students learned an isolated rule for changing a percent to a fraction but did not connect that rule to a broader understanding of percent, other mathematical ideas, or their own experiences.

▪ *Language*

How do kids think of math terms that have a colloquial use like "a hundred percent"? How could one capitalize on that knowledge? How does the colloquial use interfere with students' understanding of the mathematical concepts? Note that kids who get 100% on a test do not think of getting "1" on the test. However, "getting a hundred" on a test could make sense to a student.

■ *Capitalizing on student ideas*

Notice that the teacher essentially told the students how to solve the problem rather than how to capitalize on Eric's idea that there might only be 15 or 20. What message does this send to students?

Suggested Reading

Allinger, G. D., and J. N. Payne. 1986. "Estimation and Mental Arithmetic with Percent." In *Estimation and Mental Computation*, edited by H. L. Schoen and M. J. Zweng, 141–155. Reston, VA: The National Council of Teachers of Mathematics.

Making Sense or Memorizing Rules?

Point Seven plus Point Four Is Point Eleven

This case gives teachers an opportunity to talk about the use of concrete materials to teach decimals. It also helps them understand that manipulatives aren't "magic." By understanding the limitations of different materials, they can better anticipate the pitfalls of a particular manipulative and make informed decisions about which to use. The case also raises important issues about understanding and rote procedures.

Suggested Materials: Play money, grid paper, base-10 blocks.

Questions and Issues Raised by the Case

■ *Reasons for the error*

Did the error in the case indicate that Troy, J. J., and Melissa did not understand the problem, or was the error intrinsic to this particular problem (.7 + .4)? If the students understood the problem, why couldn't they detect the error? Why were they able to get all of the problems on the first page of the case correct but not .7 + .4? Is it possible that the students simply had a mental picture of 7 tenths (ΔΔΔΔΔΔΔ) and 4 tenths (ΔΔΔΔ) and then counted 11 tenths altogether? Would this approach work for some examples and not others? What is wrong with this approach?

Would students make these errors if they invented their own procedures? Would students make the same mistakes if the exercises were part of a real-world problem or investigation?

■ *Rote procedures, mental math, or calculator?*

Why is it that once students routinize a procedure they often do not access or apply their understandings? Automatization can be useful if it frees up mental effort to work on other worthwhile tasks, but why not use mental math for simple decimal calculations and a calculator for harder ones?

■ *Limitations of money as a model*

Teachers discussing this case often agree that money is the best model for helping students learn about decimals, because it relates so well to what they already know. But when asked to think about some of the limitations of using money, teachers can generate a fairly large list. For example, in this case the teacher used 7 dimes to represent .7, yet the conventional way of writing 7 dimes is .70, which is really 70 hundredths. Also, 7 dimes is likely to be thought of as 7 "whole things" rather than "part of a whole dollar." Likewise, students are accustomed to thinking of $1.10 as 1 dollar and 10 cents, instead of 1 dollar and 1 tenth or 10 hundredths of a dollar.

In money, the relationship between the part and the whole is abstract, compared to other models where the part is physically a piece of the whole. In other words, a dollar would not be cut into 100 parts to show the hundredths. Also, money is limited to two decimal places (or hundredths). Another model might better represent tenths or thousandths.

Students almost have to unlearn what they already know about money in order to connect it to decimals. They also confuse the names for the coins with the decimal amounts. One teacher related how a student was convinced that a nickel was a fifth of a dollar since a dime is a tenth of a dollar.

▪ *Limitations of other place-value models*

Other models considered by teachers lead to different difficulties. Base-10 blocks, for example, do provide a concrete physical model for decimals since the volume of the blocks relates directly to the place values—10 of the blocks representing tenths have the same volume as the block representing 1 whole. Yet if students have used the blocks before to represent whole numbers, they sometimes get confused about which block is which. It is valuable, however, for students to learn that the whole is arbitrary and that the other blocks can be named by figuring out their relationship to the whole. Nevertheless, for many students the blocks are abstract and do not relate well to their experiences.

Other models, such as colored chips on pegs or beans in cups, also have limitations. Students often have difficulty maintaining the whole in their minds, often thinking that the individual chips or beans are the wholes.

Metric measurement is appealing because it is a real-life application of decimals. Students who are immigrants from countries that use metric will relate especially well to this model. The part-whole confusion can still be a problem, however, since centimeters are both one hundredth of a meter and wholes in themselves. One benefit of using metric measurement is that, as students estimate and name lengths, they begin to develop the idea of relative magnitude of the decimal. In other words, they are able visually to compare 1.5 meters with 0.35 meters or 3.5 meters, and they can develop a number sense of what these decimals mean.

▪ *Is a tenth always one out of ten parts?*

What is the actual meaning of 0.3? Does it always mean 3 out of 10 parts? Could it also mean 6 out of 20 parts? When you divide 15 by 50 on the calculator, for example, you get 0.3, which is a ratio equal to $\frac{15}{50}$. This may be why decimals are so difficult to represent. Decimals are similar to percent in that they often represent a ratio other than their namesake, but usually it is the namesake that we show with models.

▪ *Connecting fractions and decimals*

The teacher in this case used fractions to help students understand decimals. Yet many textbooks present decimals before fractions. Why would this be done? Should it be done? What are the advantages and disadvantages of each approach? Is there an alternative?

Suggested Reading

Hiebert, J. 1984. "Children's Mathematical Learning: The Struggle to Link Form and Understanding." *The Elementary School Journal* 84(5): 497–513.

Hiebert, J. 1992. "Mathematical, Cognitive, and Instructional Analyses of Decimal Fractions." In *Analysis of Arithmetic for Mathematics Teaching,* edited by G. Leinhardt, R. Putnam, and R. A. Hattrup, 283–322. Hillsdale, NJ: Lawrence Erlbaum Associates.

Janvier, C. 1990. "Contextualization and Mathematics for All." In *Teaching and Learning Mathematics in the 1990s,* edited by T. J. Cooney and C. R. Hirsch, 183–193. Reston, VA: The National Council of Teachers of Mathematics.

Schielack, J. F. 1991. "Reaching Young Pupils with Technology." *Arithmetic Teacher* 38(6): 51–55.

I Still Don't See Why My Way Doesn't Work

This case has several overlapping issues with the case, "There's No One-Half Here." Both cases are about multiplying fractions. In this case, a student invented an algorithm that didn't work and wanted to know why it didn't work. The teacher spent a lot of time trying to explain an algorithm that did work, leaving the student frustrated at the end because he still did not see the flaws in his own procedure.

Questions and Issues Raised by the Case

▪ *Whose way?*

It is not unusual for discussions about this case to focus—at least in the beginning—on why the teacher's way works, not why Glen's way doesn't work. It could be that discussion participants see little value in helping someone figure out why something doesn't work—after all, why spend time thinking about something that doesn't work? Yet, knowing why something doesn't work may actually be a part of understanding why something does work. It may also be essential in avoiding future use of the erroneous method. Besides, shouldn't one value Glen's curiosity? The fact that he came back after school to figure this out was an indication of how important it was to him.

▪ *Start with a simpler problem*

One approach to handling this case would have been to help Glen see why his method didn't work. Participants may overlook the fact that Glen was already convinced that his way didn't work. Although participants may feel they are pursuing an idea that is not directly relevant to the issue raised in the case, they may need to convince themselves that Glen's method doesn't work before they can go on. They may not have had an opportunity to think about how a multiplication algorithm works and need time to explore this for themselves.

To show Glen that his method didn't work, one could have given him a simpler example, such as $1\frac{1}{2} \times 2\frac{1}{2}$, which can be solved mentally. By interpreting this problem as $1\frac{1}{2}$ of $2\frac{1}{2}$, one can understand it as $2\frac{1}{2}$ plus half of $2\frac{1}{2}$, or $1\frac{1}{4}$. So, $1\frac{1}{2}$ of $2\frac{1}{2}$ would be $3\frac{3}{4}$ all together. (See the diagram below.) If one used Glen's method of multiplying, however, one would get 2×1 and $\frac{1}{2} \times \frac{1}{2}$, which is $2\frac{1}{4}$. Therefore, Glen would probably have seen that his method was wrong.

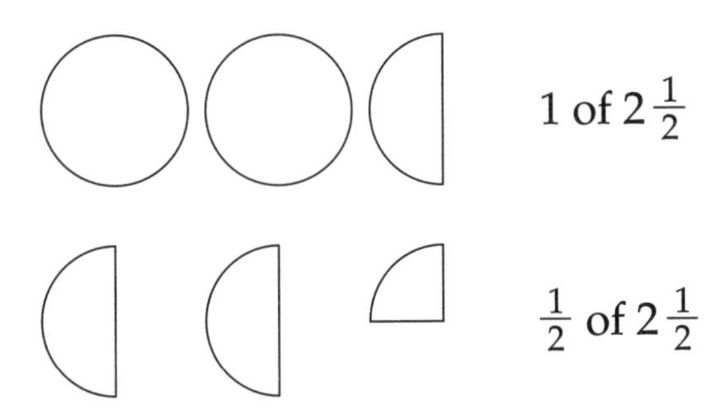

■ *An array model*

What does it mean to multiply $6\frac{3}{4} \times 5\frac{1}{3}$? Drawing on what we know about multiplying fractions, we could think of this as the area of a rectangular array. One could interpret this problem as finding the area of a rectangle that is $6\frac{3}{4}$ units long and $5\frac{1}{3}$ units wide.

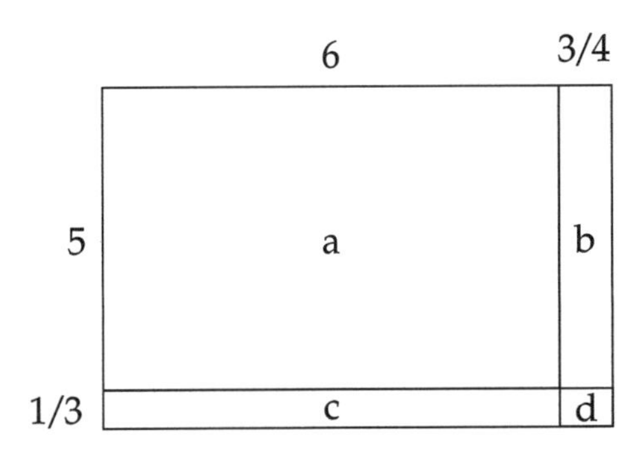

Notice that the area of section *a* is represented by 6×5, that the area of section *b* is $\frac{3}{4} \times 5$, that section *c* is $\frac{1}{3} \times 6$, and that section *d* is $\frac{1}{3} \times \frac{3}{4}$. To understand why Glen's way didn't work, we can examine the sections that he included in this algorithm: 6×5 and $\frac{1}{3} \times \frac{3}{4}$. He left out sections *b* and *c*.

■ *Relate to multiplying with whole numbers*

Another way to have helped Glen see that his method didn't work would be to relate the fraction algorithm to the whole-number algorithm. Glen might have been convinced that the whole-number algorithm made sense and that the rules for working with place value could be applied to mixed fractions. One way to see the parallels between the two algorithms is to write the fraction problem in the same format as a whole number problem.

$$\begin{array}{r} 6\frac{3}{4} \\ \times\, 5\frac{1}{3} \\ \hline \end{array} \qquad\qquad \begin{array}{r} 24 \\ \times\, 15 \\ \hline \end{array}$$

Glen incorrectly multiplied 6×5 and $\frac{3}{4} \times \frac{1}{3}$ to find the product of $6\frac{3}{4} \times 5\frac{1}{3}$. This would be similar to multiplying 2×1 and 4×5 to find the product of 24 and 15. Can you find the product of $6\frac{3}{4}$ and $5\frac{1}{3}$ using the same approach you use for whole numbers? Why is this so? What are the similarities between mixed numbers and place-value notation?

▪ *Word-problem representation*

Could a real-world problem be helpful in understanding the problem in this case? What seems like a excellent suggestion, however, points out how "unreal" the problem really is. As participants struggle to find a word problem that fits the example $6\frac{3}{4} \times 5\frac{1}{3}$, they will find that while it can be done the problem will probably be pretty contrived. Participants may be left wondering why multiplying mixed numbers is taught at all, or how much time should be spent on this algorithm.

▪ *Developmental and curricular issues*

Again, participants must examine the choices they make in the classroom. How much time should they ask their students to spend multiplying mixed numbers? Is there a place in the curriculum for learning about algorithms such as this, even though they may not be practical in daily life? Is this algorithm necessary for higher mathematics?

What do students need to know about multiplying fractions? It may be tempting for participants to throw out all multiplication involving fractions. But before this idea is confirmed, ask them to consider the consequences. Should students be able to multiply a whole number by a fraction, such as $\frac{1}{3} \times 24$? Do they also need to know how to multiply a fraction by a fraction, such as $\frac{1}{2} \times \frac{1}{4}$? What about a fraction times a mixed number, such as $\frac{1}{2} \times 3\frac{1}{2}$? Do they need to learn an algorithm, or is it enough to find the products mentally?

Suggested Reading

Janvier, C. 1990. "Contextualization and Mathematics for All." In *Teaching and Learning Mathematics in the 1990s*, edited by T. J. Cooney and C. R. Hirsch, 183–193. Reston, VA: The National Council of Teachers of Mathematics.

The Decimal Wall

In this case, the teacher tried to help her students understand and learn how to round decimal numbers. Even after using concrete materials and doing many number-sense activities with the students, they still made the typical errors. The missing connection between concrete and abstract experiences should be a major focus of the discussion.

Suggested Materials: Play money, base-10 blocks, calculators.

Questions and Issues Raised by the Case

■ *Number sense*

The teacher in this case had done several activities to build her students' number sense. She believed that her students had developed flexible ways of thinking about decimals and their relative magnitudes. Why didn't the students apply this number sense to decimal rounding? Didn't they realize that 5.810 was an unreasonable answer? What might have helped? Perhaps students could benefit with experiences related to number lines.

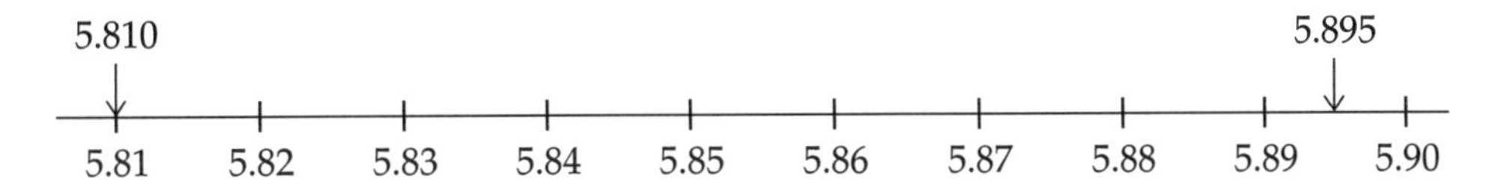

Perhaps students really did understand but were seduced into this common error and failed to realize that their answer was unreasonable. After all, you do round up and the next number after 9 is 10. Younger students make the same error when they first learn to round whole numbers. When rounding 5,895 to the nearest ten, they may write 5,810 and fail to realize that the answer isn't reasonable. These students may just need more number-sense and rounding experiences to internalize the meanings of the written symbols and recognize unreasonable answers.

■ *Relating to whole number place value—help or hindrance?*

The transition from understanding whole-number place value to decimal place value may be more difficult that it appears on the surface. For example, students know that with whole numbers the place values increase as they go from right to left, beginning with the ones, then the tens, then the hundreds, and so on. Expecting symmetry around the decimal point, students have difficulty when the decimal place values decrease as they go from the tenths, to the hundredths, and to the thousandths. In other words, it is counterintuitive that tenths have a greater value than hundredths, while tens have a lesser value than hundreds. It is also difficult for students to understand why tens are the second place to the left of the decimal point, while tenths are the first place to the right of the decimal point.

▪ *Modeling rounding*

How do you model the rounding process with base-10 blocks? What are the advantages and limitations of this model? Another model, for example, that might be considered is metric measurement, since it may help students visualize a number line.

▪ *Connecting concrete and abstract*

The students in this case were able to round numbers using concrete materials but had difficulty when they tried to apply the procedure abstractly. Why might this have been? What could have been done to make a better connection? We don't know how much experience the students had verbalizing or visualizing their work with concrete materials. Often, the materials can be handled in such a way that students go through the motions without processing the experience by talking about it. It may also help to have students draw pictures to show the rounding process.

Suggested Reading

Baroody, A. J. 1989. "Manipulatives Don't Come with Guarantees." *Arithmetic Teacher* 37(2): 4–5.

Knocking Off Zeros

In this case, sixth grade students discovered the rule for "knocking off zeros" while using base-10 blocks to represent decimal fractions. While the students seemingly understood this shortcut and the underlying fraction concepts at the time, some student work that was generated a month later depicted students' incorrect usage of the shortcut. The teacher was puzzled over why the students had forgotten "all they had learned about equivalent fractions and simplifying fractions" over the course of a month.

Suggested Materials: Base-10 blocks, fraction pieces.

Questions and Issues Raised by the Case

■ *Should the shortcut "knocking off zeros" be used?*

One of the tensions in this case is whether or not encouraging shortcuts is a good idea.

Some teachers may argue that the use of shortcuts is positive, since it often minimizes the number of steps and therefore the number of careless errors one can make in a mathematical solution. Also, shortcuts enable one to spend less time solving a problem. The use of shortcuts, however, can be problematic if the user is unclear about why and under what conditions the shortcut applies. For example, in using the knocking-off-zeros shortcut to cancel, there must be zeros in both numerator and denominator. This was either not clarified in the original lesson or not remembered a month later. Is there a danger that shortcuts can become an end in themselves, to be applied without discrimination or understanding? In this case, it appears that Tom didn't know when to use the shortcut. It seems that he had overgeneralized the application of the shortcut. Why might this happen?

■ *Why does knocking off zeros work?*

How do the base-10 blocks (e.g., flats, longs, and units) help students discover the process of knocking off zeros? By comparing the size of the blocks, students can see that pairs of fractions like $\frac{20}{100}$ and $\frac{2}{10}$ are the same. Students encounter lots of examples where both the numerator and denominator are multiples of 10, especially when exploring decimal equivalencies. This leads to a generalization that they can knock off zeros.

Why is it mathematically permissible to knock off zeros? Students could explore how to knock off zeros by dividing the numerator and denominator by a common factor. Why does this work? Some students may be ready to grasp the idea that dividing both the numerator and denominator by 10 is the same as dividing by 1. Others may notice that it is the same as reducing fractions, but they may not fully understand the mathematics underlying this procedure.

■ *Under what conditions does knocking off zeros work?*

From the examples listed by the teacher, it is not clear that students were given examples of fractions that didn't have a zero in both the numerator and denominator. For example, did students encounter examples such as $\frac{15}{100}$ or $\frac{10}{36}$ in their explorations with the base-10 blocks? In these two fractions, the numerator and denominator are not both divisible by 10, so the knocking-off-zeros shortcut is not applicable. If students don't have opportunities to apply their strategy to counterexamples, they receive no feedback on the weakness of their strategy.

■ *What was Tom thinking?*

What was the process that Tom was using to knock off zeros? In what ways could the shortcut of knocking off zeros encourage students to use this rule? Should the teacher have gone back to the manipulatives when Tom and others demonstrated this misunderstanding? If so, what should the teacher have done with the manipulatives? If not, why not? What other strategies might the teacher have used to help Tom clarify his thinking?

■ *Does using a grid with 100 squares enhance or inhibit understanding?*

In having students design a floor plan for a house on a 10-by-10 grid, the teacher asked them to figure out what part of the house was represented by each room, in terms of both a simplified fraction and a percent. What are the benefits and limitations of introducing these concepts by using a grid with 100 squares?

Notice that the denominators of all the fractions representing the area of each room on the grid were 100. Holding the denominators constant like this could have resulted in students attending only to the numerator and thus losing sight of the fact that the fraction represented a relationship between the numerator and the denominator.

On the other hand, the use of 100 as a denominator did simplify the task for students, which in turn may have enabled them to understand the relationship between fractions and percents by spending less time bogged down in computing the percent. Also, the teacher in this case may have been most interested in developing an intuitive idea of percent and fraction relationships by representing them visually as areas on a grid.

- ***When should the relationship between percents and fractions be introduced?***

Should the relationship of percents and fractions (especially fractions with denominators that are not multiples of 10) be taught together or separately in the beginning? This goes back to the issue of presenting the complexity of mathematics versus making it simple enough for students to understand. If a teacher decides to conceal some of the complexity of the concept, will students obtain a false sense of understanding that later interferes with their ability to grasp the complexities of that concept? For some teachers, it is rational to think that students should have separate experiences with ratios, fractions, simplification of fractions, and percents before combining all the concepts. However, they also need to realize that combining the concepts provides students with a richer understanding of the meaning of each and may prevent them from forming understandings that are incomplete or that apply only in special situations.

Suggested Reading

Baroody, A. J. 1989. "Manipulatives Don't Come with Guarantees." *Arithmetic Teacher* 37(2): 4–5.

Lining Up the Decimal

One rule commonly given to students learning to add decimals is to "line up the decimal points." Students who learn this way often run into problems when one of the numbers has no decimal point. The case discusses what underlies this error: Is it more than just forgetting a rule?

Suggested Materials: Play money, base-10 blocks, decimal place-value squares, calculators.

Questions and Issues Raised by the Case

- ***Number sense or more rules?***

What is at the root of the problem? Hector and the other students may have known the rule for lining up the decimals but not been able to give meaning to decimal numbers. Many teachers respond to this problem by giving students another rule to follow: "When a number doesn't have a decimal, you can put a decimal point to the right of the number." Does this solve the problem? Students in this case still don't know what .56 or 5.6 means. As long as students can't tell that .56 is less than a whole or 5.6 means 5 wholes and a little more than $\frac{1}{2}$, what is the point of having them compute with decimals? They first need to develop a number sense involving decimals. How might this be done?

- ***Number sense or place value?***

How is number sense different from place value? Is it possible to understand place value and not have number sense, or vice versa? A common approach to teaching place value is to focus on the meaning of the individual places. For example, students learn to name the value of each place and then learn that "10 hundredths is the same as one tenth," or "10 tenths is the same as 1 whole." With this approach, students may learn something about the relationship between the values of the individual places, but they often fail to grasp a deep meaning of the number as a whole. This deep and flexible understanding is regarded as number sense.

- ***Money as a model***

Although the teacher had introduced decimal addition by having students solve word problems and play store, the students apparently were not able to translate what they knew about adding with money to other situations. Did they need more experience with models other than money? It is possible that students think of money as whole dollars and whole cents, rather than whole dollars and parts of a dollar? The part-whole relationship is key to the understanding of decimals. What other models might be better for reaching this goal?

■ *Calculator*

Do students need to know this algorithm since calculators are so readily available? Would the students in this case understand whether or not their answers were reasonable? How do you help students gain this understanding?

■ *Algorithm versus concept*

What is the difference between an algorithm and a concept? Some teachers believe that they are teaching a concept if their students are using manipulatives. The teacher in this case gave her students hands-on, problem-solving experiences to help them learn the concepts underlying adding decimals, but their questions revealed that they lacked even a basic understanding of the meaning of the decimal. What should this teacher do?

Suggested Reading

Borasi, R. 1990. "The Invisible Hand Operating in Mathematics Instruction: Students' Conceptions and Expectations." In *Teaching and Learning Mathematics in the 1990s,* edited by T. J. Cooney and C. R. Hirsch, 174–182. Reston, VA: The National Council of Teachers of Mathematics.

Hiebert, J. 1984. "Children's Mathematical Learning: The Struggle to Link Form and Understanding." *The Elementary School Journal* 84(5): 497–513.

Hiebert, J. 1990. "The Role of Routine Procedures in the Development of Mathematical Competence." In *Teaching and Learning Mathematics in the 1990s,* edited by T. J. Cooney and C. R. Hirsch, 31–40. Reston, VA: The National Council of Teachers of Mathematics.

Schielack, J. F. 1991. "Reaching Young Pupils with Technology." *Arithmetic Teacher* 38(6): 51–55.

A Proportion Puzzle

Many elementary school teachers do not feel confident about their own understandings of ratio and proportion, and they may experience anxiety about this case. The discussion will be most productive if all participants have an opportunity to develop their own understandings of the mathematics before talking about the teaching in the case.

Suggested Materials: Counters, calculators.

Questions and Issues Raised by the Case

▪ *What is a proportion?*

A proportion is simply a pair of two equal ratios. One group of teachers likened it to the analogy: *hand is to arm as foot is to leg.* Similarly, *2 is to 3 as 4 is to 6;* the numerators and denominators of both ratios are related in the same way.

▪ *Efficient tool or multiple solutions?*

At the end of the case, the teacher asked himself what he wanted most—an efficient tool for solving proportions or a sense that there were multiple ways to solve a problem. This tension is often experienced by teachers who don't want to discourage student-invented solutions but also feel responsible for helping students learn an "efficient" algorithm that works in all cases. Is it possible to encourage both? Does one preclude the other? Can students be involved in testing their own invented algorithms to see if they work and are efficient for all cases? Do all students need to have the same efficient algorithm? Do they need to use the same algorithm for every problem?

The teacher states that he made up relatively simple proportion problems so that students could solve them intuitively. However, if the teacher's goal was to develop an efficient paper-and-pencil algorithm, did he also provide some problems that could not be solved so easily mentally? Otherwise, the mental approach is the more efficient and more natural approach. Why wouldn't students balk at using the less efficient pencil-and-paper algorithm?

▪ *Building on student thinking*

Students enter school with the capability of figuring out mathematical problems for themselves. By the end of first grade, however, students are already learning not to depend on their own reasoning to solve mathematical problems and their own judgment to evaluate whether or not something is mathematically correct. Instead, the source of knowing comes to reside in textbooks or teachers.

The teacher in this case might have asked his students to share their thinking with the rest of the class and challenged the class to decide if the methods were valid for all problems. Or, as the teacher suggested, the students might have been asked to figure out if the methods were equivalent. Is the amount of time needed to carry out discussions like this worth it? What about students who resist this approach and demand to just be told how to get the answer? Will students just get befuddled by listening to each other's explanations?

Calculator

Does the calculator method suggested by Chris work? Is the paper-and-pencil method necessary if students can do simple problems mentally and use the calculator for more difficult problems?

Messy proportions

Some proportions are easy to solve because the numbers in the proportion are related by being multiples of each other. The "whale problem" in the case is like this. The "bird problem" is a little more difficult, since 20 is not a multiple of 30 and 30 is not a multiple of 45. But both of these problems are relatively easy, partly because they have whole-number solutions. Many real-life problems, however, do not, and students are seldom asked to solve this kind of problem. What are the positive and negative consequences of not giving students messy problems to solve?

Suggested Reading

Cramer, K., and T. Post. 1993. "Making Connections: A Case for Proportionality." *Arithmetic Teacher* 40(6): 342–346.

Curcio, F. R. 1990. "Mathematics as Communication: Using a Language-Experience Approach in the Elementary Grades." In *Teaching and Learning Mathematics in the 1990s*, edited by T. J. Cooney and C. R. Hirsch, 69–75. Reston, VA: The National Council of Teachers of Mathematics.

Teaching as Questioning

Zeros Sometimes Make a Difference

Teachers have come to believe that lack of place-value understanding is the usual problem and using manipulatives is the answer. However, closer examination of the problem and more scrutiny of how manipulatives influence student thinking reveal that neither problem nor solution is so simple. This case gives some insight into these issues. By examining commonly used manipulatives such as decimal squares, money, and base-10 blocks, the case may stimulate teachers to consider other alternatives, such as measurement and student drawings. It also brings up the idea of "benchmarks" and can lead to a discussion about number sense.

Suggested Materials: Decimal place-value squares, base-10 blocks.

Questions and Issues Raised by the Case

■ *Showing zero with place-value materials*

How do you represent the zero in place-value materials such as decimal squares or base-10 blocks? Teachers suggest that the best way to represent the zero is by using a place-value mat that has an area designated for blocks representing hundreds, tens, ones, tenths, hundredths, and thousandths. To represent 0.03, for example, one shows 3 blocks in the hundredths place; the absence of blocks in the tenths place represents the zero.

How is this explained to students? Is this just another rule to follow? Is the forest getting lost in the trees? Is the zero better represented with other materials such as decimal squares?

■ *Comparing the strengths and weaknesses of place-value materials*

How is using base-10 blocks similar or dissimilar to using decimal squares?

Decimal squares show a visual representation of the part and the whole. The square is divided into thousandths, and 235 of those thousandths are shaded. Base-10 blocks do not show this relationship. With base-10 blocks, the whole must be imagined.

Base-10 blocks are sometimes interpreted by students to be whole units—i.e., 3 tenths and 5 hundredths is thought of as 3 things and 5 things. Students can say and write the number without understanding the part-whole relationship.

Could metric measurement be a useful representation? It is especially relevant to students from countries that use the metric system. One disadvantage is that the parts can be interpreted as wholes and the part-whole relationship confused. However, measurement seems to build a better sense of relative magnitude, since the decimal number is associated with a physical length, weight, or capacity, if real materials are used.

Is this approach accessible and relevant to students of diverse cultures? Decimal squares and base-10 blocks are universally accessible and metric measurement might have more relevance to many recent immigrants.

Decimal squares, base-10 blocks, and meter sticks are all predivided. Is this an advantage or a disadvantage? What experiences could we provide that would not use predivided materials?

▪ *Relative magnitude*

What is meant by relative magnitude? If students know which is more, five tenths of something or three hundredths of something, they have a sense of the relative magnitude of these two numbers. Relative magnitude is closely related to number sense and place-value concepts. Why might it be important for students to be able to compare decimals?

▪ *Benchmarks*

Benchmarks, or quantities that are familiar and accessible, help students judge the relative magnitude of numbers. Benchmarks such as $\frac{1}{2}$ or 0.50, for example, help one visualize and interpret 0.47 as a little less than $\frac{1}{2}$. Woody may have lacked benchmarks for interpreting decimals. How can decimal squares or base-10 blocks help develop benchmarks? Estimation strategies that rely on benchmarks can be a powerful vehicle for building a sense of relative magnitude.

▪ *Interference of prior learning with new learning*

Students can become confused because the place-values do not correspond in a logical way for them. For example, although tens are less than hundreds, tenths are more than hundredths.

▪ *Zeros do make a difference*

Zeros do make a difference, whether they are on the right or the left. Note that 0.35 equals 35 hundredths and 0.350 equals 350 thousandths.

■ *Relevancy or rules?*

Is it possible that Woody and other students were missing the relevancy of decimals to their lives? What materials or experiences would have helped make decimals more relevant? Could we think of an approach that would be more holistic? What might be the equivalent of a "whole-language" approach?

Suggested Reading

Cuevas, G. 1990. "Increasing the Achievement and Participation of Language Minority Students in Mathematics Education." In *Teaching and Learning Mathematics in the 1990s,* edited by T. J. Cooney and C. R. Hirsch, 159–165. Reston, VA: The National Council of Teachers of Mathematics.

Long, M. J., and M. Ben-Hur. 1991. "Informing Learning Through the Clinical Interview." *Arithmetic Teacher* 38(6): 44–46.

Webb, N., and D. Briars. 1990. "Assessment in Mathematics Classrooms, K–8." In *Teaching and Learning Mathematics in the 1990s,* edited by T. J. Cooney and C. R. Hirsch, 108–117. Reston, VA: The National Council of Teachers of Mathematics.

Six-Tenths or Four-Fifths of a Dollar?

Students' journal writings can lead to insights about student thinking and how that relates to our teaching. This case presents a sequence of lessons using fraction kits created by the students in the class. Although the fraction kit provided many hands-on experiences, questions are raised about how it was used and its transferability to the journal task.

Suggested Materials: Fraction pieces, counters.

Questions and Issues Raised by the Case

▪ *Understanding children's thinking*

In what ways did the students show rational ways of thinking about the comparison of $\frac{6}{10}$ and $\frac{4}{5}$? Chris seemed to have a clear idea of fraction equivalency and was able to illustrate that understanding in a drawing. On the other hand, students like Cindy (who thought that $\frac{6}{10}$ was 2 more than $\frac{4}{5}$) may have only focused on part of the fraction (that the numerator of $\frac{6}{10}$ was 2 more than the numerator of $\frac{4}{5}$). Like many students, she did not consider the relationship between the numerator and the denominator. Why might students focus only on the numerator? What makes this a reasonable thing to do? In what ways could students be led to focus on the numerator exclusively? (Note that questions like, "How many eighths would be equal to $\frac{1}{4}$?" can be answered with "2.")

How might one interpret Nikki's drawing? Some see it as representing 6 tenths and 4 fifths, where both the tenths and fifths are represented as little circles. Again, Nikki may have been focusing primarily on the numerators and ignoring the denominators. Others think that Nikki's drawing represents 6 out of 10, since she drew 10 circles and circled 6.

▪ *Complexity of the journal task*

The teacher in the case asked whether the students would have responded differently if the question had been, "Would you rather have $\frac{3}{4}$ or $\frac{5}{8}$ of a chocolate bar?" How would students go about showing $\frac{5}{8}$ of a dollar? How would it be different than showing $\frac{5}{8}$ of a chocolate bar? This brings up the difference in representing parts of a set (100 cents or a dollar) and parts of a whole region (a square or circle in a fraction kit). Is one more difficult than the other?

- ***Limited English Proficiency and journal writing***

Educators of both immigrant students with limited English proficiency and students for whom writing is a chore may shy away from having their students write in journals. In what ways are these concerns legitimate? How might these concerns be addressed? Should students be encouraged to write in their primary language? What if the teacher cannot read the student's writing? How might one help all students understand the value of writing so that they are more motivated to participate?

- ***Internalizing fraction concepts***

What are the strengths of the fraction kit in promoting understanding of fraction concepts? Most will agree that one reason fraction kits are valuable is that students experience cutting "whole" squares, rectangles, or circles of different colors into parts. Also, the pieces are easy to visualize and can be used to solve simple problems such as, "How many sixteenths are equal to one half?" They help students develop the language for fractions informally and become familiar with basic concepts such as equivalency.

Are there limitations of fraction kits? Even though students have the experience of making their own fraction kits, they may lose sight of the significance of the part-whole relationship. Once the sixths, eighths, or thirds are cut apart and intermingled, they may just become "puzzle pieces" to be used to solve problems. Students may perform the actions of "fitting two eighths on a fourth" but not be thinking of how these fractions relate to each other or to the whole. Even though they may say the names of the pieces—"two-eighths" or "one-fourth"—is it possible that names may carry no more meaning than "two blues" or "one red?" Furthermore, it is possible that showing that $\frac{2}{8} = \frac{1}{4}$ requires no more thought than remembering the names of the blue and red pieces and counting how many blues fit on a red.

Are there other kinds of experiences or questions that might provoke a deeper understanding of the part-whole relationships represented by fractions? Consider the value of partitioning wholes into parts. Experiences in which the students have to figure out the relationship between fractions by drawing or making their own models may be at the heart of understanding part-whole relationships. The fraction kit provides some of this experience, but perhaps it needs greater emphasis.

Suggested Reading

Behr, M. J., T. R. Post, and I. Wachsmuth. 1986. "Estimation and Children's Concept of Rational Number Size." In *Estimation and Mental Computation,* edited by H. L. Schoen and M. J. Zweng, 103–111. Reston, VA: The National Council of Teachers of Mathematics.

Cuevas, G. 1990. "Increasing the Achievement and Participation of Language Minority Students in Mathematics Education." In *Teaching and Learning Mathematics in the 1990s,* edited by T. J. Cooney and C. R. Hirsch, 159–165. Reston, VA: The National Council of Teachers of Mathematics.

Testing Theories

The teaching approach in this case is based on the belief that students learn by processing information and by formulating and testing their own ideas. Students are viewed here as rational decision makers, capable of determining for themselves what is right or wrong. Discussion participants can discuss the value of this approach and how well it is carried out in this lesson.

Suggested Materials: Grid paper, transparency of grid, base-10 blocks, calculators.

Questions and Issues Raised by the Case

▪ *Testing theories*

The teacher in this case spent a lot of time allowing students to present and test alternative theories for multiplying decimals. Why didn't she just give them the rule? Wouldn't it have been quicker? Wasn't there a danger that students would remember the wrong theory?

Thinking about the reasons the teacher chose to take this route is a valuable endeavor. Was her purpose simply to teach the algorithm or was she trying to communicate something else to her students? Perhaps she wanted her students to understand that mathematics reasoning was something they could do, that it was not a magical nor an incomprehensible recipe. Maybe she wanted her students to be able to reinvent the algorithm if they forgot it, or she may have wanted them to know that there were many ways to think about a mathematical problem. She could also have been modeling the way mathematicians and others develop and test theories. Finally, her purpose might have been to foster a risk-taking disposition in her students, so they would share and be open to debate about their theories.

▪ *Understanding multiplication*

Many of us file mathematical ideas in separate categories, failing to recognize the powerful connections that weave throughout mathematics. Whole numbers are filed separately from decimals, multiplication is filed separately from division, and measurement is filed separately from numbers. The challenge for teachers is to help students build a cross-referencing system for all mathematical ideas. The stronger these connections, the more meaningful mathematics will be to students.

In this case, the teacher used the array to show how multiplying decimals related to multiplying whole numbers. In this way, she built a deeper understanding of the meaning of multiplication, one that could later be extended to fractions and even algebra.

■ *Is the length tenths or hundredths?*

Teachers who have used arrays to help students understand multiplication of decimals often run into a snag. Students get confused about why the width of the rectangle is 0.5 and not .05, or why the length is 1.2 and not 0.12. After all, don't you find the width by counting 5 squares across and isn't each square one hundredth, thus 0.05? It is easy to confuse the two-dimensional area of the small squares (0.01 square units) with the one-dimensional length of the sides of these squares (0.1 unit).

■ *60 out of 100 squares?*

Another potential trouble spot in the lesson was when the teacher assumed students understood that the area of the rectangle on the grid was 60 out of 100. Why wasn't it 60 out of 200? Perhaps it would have been better to talk about the value of 1 square (1 out of 100 or 0.01), and then to reason that 60 squares would be 0.60.

Suggested Reading

Damarin, S. K. 1990. "Teaching Mathematics: A Feminist Perspective." In *Teaching and Learning Mathematics in the 1990s*, edited by T. J. Cooney and C. R. Hirsch, 144–151. Reston, VA: The National Council of Teachers of Mathematics.

Hiebert, J. 1992. "Mathematical, Cognitive, and Instructional Analyses of Decimal Fractions." In *Analysis of Arithmetic for Mathematics Teaching*, edited by G. Leinhardt, R. Putnam, and R. A. Hattrup, 283–322. Hillsdale, NJ: Lawrence Erlbaum Associates.

Stigler, J. W. 1988. "Research into Practice: The Use of Verbal Explanation in Japanese and American Classrooms." *Arithmetic Teacher* 36(2): 27–29.

Stigler, J. W., and H. W. Stevenson. 1991. "How Asian Teachers Polish Each Lesson to Perfection." *American Educator* 15(1): 12–47.

Webb, N., and D. Briars. 1990. "Assessment in Mathematics Classrooms, K–8." In *Teaching and Learning Mathematics in the 1990s*, edited by T. J. Cooney and C. R. Hirsch, 108–117. Reston, VA: The National Council of Teachers of Mathematics.

Hugh's Invention

This case frames an interaction between a teacher and a student engaged in a measurement investigation. The teacher, puzzled by an answer that the student had written on his paper, probed his understanding. Within this interaction, the teacher discovered that the student, Hugh, actually had a sophisticated understanding of fractions and ratios. Discussion participants may discuss how it was possible for Hugh to develop the insights that he had, and what type of teaching enables teachers to discover whether students have developed sophisticated thinking like Hugh's.

Questions and Issues Raised by the Case

■ *Developing number sense*

Hugh demonstrated insight into the relationship between whole numbers and fractions and ratios. He also displayed keen visual perception of fractional amounts. How did Hugh develop this insight? How many other students shared Hugh's insights? Hugh seemed to understand the idea that the closer the numerator and the denominator were to each other, the closer the fraction was to 1. What kinds of activities enable students to develop these types of insights?

■ *Estimating equivalent fractions*

In measuring and reporting how far he jumped, Hugh gave the answer of $130\frac{29}{31}$ centimeters. When the teacher asked him about the meaning of $\frac{29}{31}$, he responded that it meant that his jump was really close to 131. Why did Hugh choose 31 as his denominator? It's an unusual number for a student to come up with. It is interesting that the denominator was the same as the last two digits of the whole number 131. Was this why Hugh selected 31 as the denominator?

When the teacher probed Hugh about his reason for selecting 31, he responded that he could have used any number he wanted—it could have been 31, 15, or 78. When asked about the numerator if he had chosen 78, Hugh responded that he wouldn't have 29. Instead he would have 72, which he quickly amended to 74. In this set of interactions, Hugh was demonstrating keen visual perception and understanding of equivalent fractions: $\frac{29}{31}$ is about equal to $\frac{74}{78}$ and $\frac{14}{15}$. He showed similar perception and understanding when the teacher used the example of the thumb and straw (Note: $\frac{1}{4} = \frac{5}{20}$). What concrete experiences enabled Hugh to develop this type of understanding? What was Hugh visualizing when he arrived at solutions? It is possible that Hugh was visualizing subdivisions? He seemed able to see in his mind's eye a set of subdivisions and the number of those divisions used to measure the thumb. Did Hugh develop this understanding, or was it something that he was born with? Why might the example of the thumb and straw be helpful?

▪ *Standard versus non-standard measure*

In this case, the teacher designed the investigation so that students had access to both standard and nonstandard measuring instruments. How might the use of both kinds of measuring instruments help or hinder the type of understanding demonstrated by Hugh? Should one type of instrument be used before the other? Should students be taught to measure with nonstandard measuring tools first, and then be given standard tools later because of their developmental level? What about the other way around: Should students first use the standard measuring tools, and then, after developing a visual sense of fractional amounts, switch to nonstandard tools in which they would have to use their own developed sense of the measuring units and subunits?

▪ *Establishing the right environment for students to demonstrate what they know*

Noticing Hugh's written measurement of $130\frac{29}{31}$, the teacher repeatedly came back to him in an attempt to understand his thinking. How can teachers acquire the time and ability to get better insight into their students' thinking? In part, spending more class time on student-generated questions and encouraging students either to discuss or write about their thinking could help in this effort. How can questions be devised that are thought provoking and accessible to students? How can teachers create a comfortable environment that enables risk taking?

▪ *Limiting students' natural ability*

The teacher was proactive in trying to understand Hugh's thinking. She interviewed him by asking questions which she formulated on her feet. Is it possible that teachers would discover that more of their students share Hugh's understanding if they engaged them in similar kinds of dialogue? While there are some indications that Hugh had well-developed intuitive skills, is it possible that some teaching can unintentionally hinder the development of such insightful thinking? Do traditional teaching episodes limit (or hinder) students' intuitive thinking about mathematics? Some argue that this type of thinking is a part of students' natural ability, which is often "trained out" of them. Do we limit students by our expectations of what mathematics is or even by our expectations of the students themselves? How can teachers encourage students to develop this creative, intuitive ability, and at the same time develop student understanding of more "standardized" math? At what point in students' development is it appropriate to teach these intuitive skills? Is it possible to teach intuitive skills?

■ *What would you do with a Hugh in your classroom?*

Some might want to assign a student like Hugh a special role in the classroom. Such an assignment could keep Hugh motivated and help the teacher by having another person to explain mathematics to other children. Did Hugh have the social and language skills necessary to be a tutor? There is some indication in the case that he might not have enjoyed discussing his thinking with others as much as he enjoyed the mathematical exploration.

How can teachers attend to a student like Hugh and keep him motivated to learn mathematics? If Hugh is an exceptional math student, he might be bored by teaching that addresses the others in his class.

Suggested Reading

Carter, H. L. 1986. "Linking Estimation to Psychological Variables in the Early Years." In *Estimation and Mental Computation,* edited by H. L. Schoen and M. J. Zweng, 74–81. Reston, VA: The National Council of Teachers of Mathematics.

Kamii, C. 1990. "Constructivism and Beginning Arithmetic, K–12." In *Teaching and Learning Mathematics in the 1990s,* edited by T. J. Cooney and C. R. Hirsch, 22–30. Reston, VA: The National Council of Teachers of Mathematics.

Thirteen Can't Fit Over Twelve

This case raises questions about the nature of good teaching. The author's underlying philosophy is explicitly presented as part of the case so that teachers can analyze how—or whether—his philosophical beliefs were carried out in practice. More importantly, teachers must try to understand how the author's critical-thinking approach impacted his students, and they must try to evaluate whether that impact was positive. The case is controversial, partly because the author seems somewhat arrogant and partly because so many of his practices are at odds with what is commonly considered good teaching. This case is divided two parts: the main text and the Epilogue. One way to use this case is to hand out and discuss each section separately. Dividing the discussion of this case facilitates the deepening of the discussion. When both parts are considered at once there is a danger that the discussion will miss some interesting issues that are embedded in the main text of the case.

An interesting way to begin discussion is to have each participant grade (from A to F) the teaching represented in this case. Discussion can then focus on why a particular grade was given. At the end, ask the participants if they would now change their grades.

Questions and Issues Raised by the Case

■ *Critical thinking or problem solving?*

What is critical thinking? How does it differ from problem solving? Teachers seldom associate critical thinking with mathematics, and they need to grapple with what it might mean in this context. Critical thinking entails analyzing choices and making wise or rational decisions. Although problem solving always involves critical thinking, critical thinking does not always involve solving problems. For example, one aspect of critical thinking is detecting errors or misleading information. Another is analyzing and evaluating possible solution strategies.

■ *On a tangent*

One of the more annoying aspects of this case for some teachers is that the whole lesson is a "birdwalk." Most teachers have been taught to believe that getting off the topic is bad teaching. What was the teacher's objective? Was he really on a tangent? Is it possible the author was more interested in having students experience the mathematical thinking necessary to develop a definition than in having them learn a definition?

- ***Introducing confusion***

Why would a teacher introduce a complex fraction when students haven't ever seen one before? Wouldn't it just confuse or intimidate them? What are the possible benefits? By introducing examples that pushed students to refine their definitions, the teacher modeled a critical-thinking process. The question is, Do students adopt this process by seeing it modeled, or do they need some other kind of experience?

- ***Nature of the discussion***

An analysis of the classroom discussion shows that most of the questions were asked by the teacher while answers were given by students. Was this appropriate? Was this consistent with the author's philosophy?

- ***Risk taking***

What evidence in this case indicates that students were—or were not—used to taking risks in this classroom? Why did the teacher force students to take a stand on whether or not $\frac{9}{9}$ was an improper fraction? Wasn't he setting them up to be wrong? How might this have helped or hindered their self-esteem?

Epilogue

- ***Teachers' role in facilitating critical thinking***

The Epilogue provides a rationale for the decisions and actions that the teacher made during the lesson. According to the case, he hoped to develop a "disposition toward critical thinking" in students. Given that this was the teacher's objective, was it appropriate for him to bounce back and forth among students' responses? Did the main ideas get lost because the discussion developed along tangential lines of thinking? While some may question the teachers' actions or attitudes toward students, one could also argue that in this classroom the students had the authority to determine the validity of various answers. There are many examples where students were given the opportunity to think about various answers and determine whether proposed answers made sense.

Suggested Reading

Beyer, B. 1983. "Common Sense About Teaching Skills." *Educational Leadership* 41(3): 44–49.

Borasi, R. 1990. "The Invisible Hand Operating in Mathematics Instruction: Students' Conceptions and Expectations." In *Teaching and Learning Mathematics in the 1990s*, edited by T. J. Cooney and C. R. Hirsch, 174–182. Reston, VA: The National Council of Teachers of Mathematics.

Passmore, J. 1967. "On Teaching to Be Critical." In *The Concept of Education*, edited by R. S. Peters. London: Routledge & Kegan Paul.

Reflecting

Problem of the Week: "Bounce, Bounce"

Many classroom teachers understand the value of assigning a "problem of the week." They want their students to be involved in sustained work on a problem over a period of time, and they want students to know that mathematics is more than computation. However, this strategy requires considerable dedication on the part of the teacher. This case focuses on the difficulties of implementation and asks teachers to consider how to make the process work without burning themselves out.

Questions and Issues Raised by the Case

▪ *Understanding students' error patterns*

Student solutions to this problem fell into four strategies: dividing, subtracting, adding, and "taking half." Some of the strategies, although rational, led to errors on the sixth bounce.

Derrick, for example, used the subtraction approach, subtracting $\frac{1}{2}$ of the previous height on each bounce. Why did he subtract $\frac{1}{2}$ from $\frac{1}{2}$ instead of subtracting $\frac{1}{4}$ from $\frac{1}{2}$? Perhaps he didn't think he could subtract from a number that was less than 1. Was this just a fluke, or does our instruction of fractions somehow contribute to this error? How might it be prevented?

What was the reason for Leslie's error? Could it be that many students don't think they can divide a "little" number by a "big" number? Perhaps she thought the answer was zero because "2 doesn't go into 1." How might we design our division instruction to try to prevent this common error? How can we help students become better able to catch these errors themselves?

Why was the "take half" method used by Jennifer and Donald more successful?

▪ *The challenge of motivating all students*

Aaron was a "star" last week, but this week he barely tried. How can we motivate students to make an honest effort to solve the problem each week? What general strategies can draw students into the problem and get them started?

What about the student who has very low mathematics skills? Will this necessarily inhibit his or her problem solving, or can many problems be solved through reasoning or trial and error? What about students with no confidence in their math ability?

Some things to consider: Show that students' work is valued by putting selected solutions in a portfolio, displaying interesting solutions on the bulletin board, or sending a letter home to parents. Perhaps there could be some schoolwide contest or activity that reinforces the importance of problem solving.

■ *Evaluating solutions*

What do you think of the teacher's evaluation system? Do you think students are motivated by this, or could it inhibit motivation? Might it also inhibit reflection and self-critiquing, as the teacher suggests? Consider how long it might take to evaluate thirty solutions. Is there an easier or more efficient way? What are the advantages or drawbacks of different approaches to evaluation?

■ *Debriefing solutions*

How would you discuss the students' solutions? Would you just give back their papers with written comments? How much time would you spend in discussion? Would you discuss several student solutions?

Suggested Reading

Holmes, E. E. 1990. "Motivation: An Essential Component of Mathematics Instruction." In *Teaching and Learning Mathematics in the 1990s*, edited by T. J. Cooney and C. R. Hirsch, 101–107. Reston, VA: The National Council of Teachers of Mathematics.

Webb, N., and D. Briars. 1990. "Assessment in Mathematics Classrooms, K–8." In *Teaching and Learning Mathematics in the 1990s*, edited by T. J. Cooney and C. R. Hirsch, 108–117. Reston, VA: The National Council of Teachers of Mathematics.

Function Machine

The student dialogue and journal writings in this case provide a rich example of how students can solve problems in many different ways. The case also raises issues concerning the use of journal writing with limited-English speakers.

Questions and Issues Raised by the Case

■ *Patterns and functions*

What is meant by a function? How is a function different from a pattern? In these function tables, many of the students described patterns in the numbers going down a column. Thomas' journal described such patterns. The function idea, however, is the relationship between a number in one column and the corresponding number in the other column. Given the number in the first column, one can determine the number in the other column using that relationship. Although her journal was a little unclear, Lourdes appeared to see the function and described it, saying that 5 times the number of quarters gave the number of nickels.

■ *Testing theories*

This teacher asked students to offer theories about different ways to think about function problems. Once a student in the class articulated a theory, she asked the whole class to test that theory on other problems. For example, when Julio was asked for his theory, he said that he "figured out how many people, then did times 2." The teacher called on the class to validate Julio's theory by testing it on other problems. She then called on Sam to explain his theory and proceeded to help the class determine whether or not his theory worked. Why might the teacher be calling these methods "theories"? What is a theory? In what ways was the environment of this class conducive to having students produce alternative theories? Who holds the authority for determining what solutions are valid or invalid in a "theory-testing" approach to learning? What are the implications for teaching and learning?

■ *Alternative ways to solve problems*

Many students, as well as adults, believe that in mathematics there is only one correct method and one correct answer. What beliefs are implied by the dialogue in this lesson? How do different beliefs influence student learning and performance? How does one change beliefs, especially when students are older and more entrenched in these beliefs?

■ *Journal writing and language skills*

Is journal writing appropriate for all students, including second-language learners and English-speaking students with weak language capabilities? How can the writing be structured to accommodate learners with various skills, learning styles, and backgrounds? Is journal writing worth the extra time and effort?

If the teacher is unfamiliar with a learner's primary language, how can he or she respond to journal writing if it is done in the primary language? Notice that although Rosa's writing is in Spanish, her work indicated that she might have been seeing some proportional relationships. How could one follow up on this observation?

Suggested Reading

Azzolino, A. 1990. "Writing as a Tool for Teaching Mathematics: The Silent Revolution." In *Teaching and Learning Mathematics in the 1990s,* edited by T. J. Cooney and C. R. Hirsch, 92–100. Reston, VA: The National Council of Teachers of Mathematics.

Cuevas, G. 1990. "Increasing the Achievement and Participation of Language Minority Students in Mathematics Education." In *Teaching and Learning Mathematics in the 1990s,* edited by T. J. Cooney and C. R. Hirsch, 159–165. Reston, VA: The National Council of Teachers of Mathematics.

Wilde, S. 1991. "Learning to Write About Mathematics." *Arithmetic Teacher* 38(6): 38–43.

Why Isn't It One Less?

This case is about the multiple solution strategies fourth-grade students used to solve a word problem involving proportions. While none of the students used formal knowledge of proportions, many students used informal knowledge and their understanding of multiples to find correct solutions. Other students did not recognize or attend to the multiplicative features of the problem and gave an incorrect answer of "one less."

Suggested Materials: Counters.

Questions and Issues Raised by the Case

▪ *Solutions involving multiples*

Most of the students made tables with matching multiples of 4s and 5s until they arrived at 35. Lee used multiples in a more abstract way, reasoning that since $7 \times 5 = 35$, then 7×4 or 28 would give the answer. Amy's solution demonstrated another, perhaps more sophisticated, strategy involving multiples. She said, "I know 7 times 5 is 35, so 35 is 7 groups of your books." She reasoned that in each exchange of books you got one less than the bookstore, so if the bookstore received a total of 35 books then you received 35 minus 7 comic books.

▪ *The "one-less" solution*

Several students said the solution was 34. How did they get this answer? It appears that some students, like Amy, may have reasoned that if you had 35 comic books to trade and you traded 5 books each time, you would trade a total of 7 times. However, instead of subtracting 1 group of 7 books (1 for each trade), they simply subtracted 1 from 35 to get 34 comics.

Is it possible that the students used other strategies to get the answer 34? Although the case writer did not provide examples of student work to confirm or disconfirm that this may have happened, there are other plausible strategies that could have been used. It is common for some students, especially younger students, to add instead of multiply to solve problems involving proportions. For example, they might have asked themselves, "How many comic books do I add to 5 comic books to end up with 35?" Arriving at an answer of 30, the students would then have added 30 to the 4 comic books and gotten a total of 34 comic books.

Students also may have gotten 34 as an answer by subtracting. They might have thought that since 4 comic books is 1 less than 5 comic books, they also needed to take 1 away from 35 to get 34 comic books. Again, students who have difficulty reasoning about proportions in terms of multiplication may resort to strategies that involve addition or subtraction. Some believe that students who solve proportions by addition or subtraction are not developmentally ready for proportional reasoning involving multiplicative relationships.

■ *How can the comic-book problem be solved with a proportion algorithm?*

Some teachers may say that the problem could be solved through setting up a proportion.

Based on the problem statement, one could mathematically express the comic book proportion as $\frac{4}{5} = \frac{x}{35}$. This proportion is made up of the two equal ratios, $\frac{4}{5}$ and $\frac{x}{35}$. Setting up this proportion can be a real stumbling block for students. How do you know which number goes in the numerator and which goes in the denominator? A critical component of all proportions is that they express a multiplicative relationship among quantities, so the relationship of x:35 is a multiple of 4:5. The problem could be solved by finding the number which is multiplied by 5 to get 35, and then multiplying the numerator (4) by that number (7). Teachers may also solve this problem by cross multiplying and getting $4(35) = 5x$ and then dividing both sides by 5 to get the answer.

■ *Are proportions appropriate for fourth-grade students?*

Teachers may argue that fourth-grade students are not developmentally ready to handle proportions. Does this mean that all students are not ready to handle proportions until fifth grade? During this discussion, you may want to refer to the NCTM Curriculum Standards which locate the development of reasoning about proportions in the fifth- to eighth-grade curriculums. While proportional reasoning is not listed in the curriculum for students in the first to fourth grades, it does refer to the study of patterns and relationships. The student solutions in this case illustrate that they used proportions informally through patterns and relationships. Proportional thinking may not be explicitly called for in the fourth-grade curriculum, but it may be important to lay the groundwork for future work in this area. Based on the teacher's account of student responses, it appears that most students were able to engage in proportional thinking at some level. The value of this task was that it supported multiple solution strategies appropriate for students with different developmental capabilities.

■ *What next?*

What can be done with the students who got "one less" as an answer? One possible solution is to have students talk through or act out the problem. Students could take on roles as the comic-store owner and the buyer and then actually exchange something like books, chips, or cubes. More discussion about the different strategies may help them expand their understanding.

Suggested Reading

Cramer, K., and T. Post. 1993. "Making Connections: A Case for Proportionality." *Arithmetic Teacher* 40(6): 342–346.

Manipulatives Aren't Magic

Rote Manipulatives?

This case may not be easy to discuss. Discussants may dismiss the lesson's failures as the fault of an inexperienced teacher. However, the key issue in this case is the use of manipulatives: When used to match an algorithm step by step, they can become no less rote than the paper-and-pencil procedure itself.

Suggested Materials: Counters (i.e., cubes), plastic cups.

Questions and Issues Raised by the Case

▪ *What is meant by rote manipulatives?*

Manipulatives are sometimes perceived by teachers as a panacea—if you use manipulatives, you are automatically teaching for understanding. Overlooked is the fact that manipulatives can sometimes be just as abstract and as rote as pencil-and-paper procedures. In this case, the teacher may have chosen the particular task because she already understood division and proportion and found the connection between them fascinating. Looking at Samatha's work, however, we see how confusing and futile it can seem from a student's point of view.

Even though the teacher carefully led them through the task step by step, the students must have wondered what the point was. Although one use of manipulatives is to explain the individual steps of a procedure, this requires relatively low-level thinking. In other words, do we want students to understand the steps of the procedure, or to understand that division and proportion are related, or both? In what order should these purposes be considered? How can manipulatives be used to foster higher-level thinking among students and support broader understandings?

▪ *Meaningless tasks*

In this case, the sixth-grade students who already knew how to do division may have thought this task made a simple problem difficult. In this sense, the task could have reinforced the idea that mathematics has little purpose and is something you do because the teacher asks. Prior to the instruction on proportions, Samatha probably could have divided 11 cookies evenly between 2 people without difficulty. Yet she was so caught up in illustrating the problem with the proportion procedure that she failed to realize her answer didn't make sense.

▪ *Connecting division with proportions*

Is it important to help students see the connection between division and proportions? Mathematics educators have been criticized in the past for failing to help students relate ideas within mathematics. Is it possible to relate these ideas in a more intuitive way? What if you simply asked students to solve simple proportion problems using concrete materials and then to explain how they did it? It is likely that some students would naturally use division as a solution strategy, and their work could be used as a springboard for discussion.

▪ *Teacher judgment*

Did the teacher use good judgment in deciding on this lesson for her sixth-grade students, even though it was recommended for third-grade students? Teachers with more experience—the teacher in the case only had three years of experience—may think not. But is the manipulative task appropriate or useful for third graders?

Teachers are often not comfortable enough with their own mathematics abilities to question "experts" who give presentations at conferences. Yet developing a critical eye for their work and others' recommendations is crucial. If teachers try something and it doesn't work, they may think they failed; their confidence to try other new things is weakened. However, if teachers believe in their own abilities to judge the value of an experience and learn from it, their confidence and knowledge is strengthened.

Suggested Reading

Baroody, A. J. 1989. "Manipulatives Don't Come with Guarantees." *Arithmetic Teacher* 37(2): 4–5.

Driscoll, M., and B. Lord. 1990. "Professionals in a Changing Profession." In *Teaching and Learning Mathematics in the 1990s,* edited by T. J. Cooney and C. R. Hirsch, 237–245. Reston, VA: The National Council of Teachers of Mathematics.

Hiebert, J. 1990. "The Role of Routine Procedures in the Development of Mathematical Competence." In *Teaching and Learning Mathematics in the 1990s,* edited by T. J. Cooney and C. R. Hirsch, 31–40. Reston, VA: The National Council of Teachers of Mathematics.

Two Green Triangles

This case is a reminder that communicating the purpose of a lesson to students is vital. It also suggests that breaking things down into little steps to make them easier for students may be a misguided, though well-intentioned approach. A more open-ended, problem-solving strategy could be much more powerful.

Suggested Materials: Pattern blocks.

Questions and Issues Raised by the Case

▪ *Arbitrarily designating the whole*

What would students say was the point of this lesson? Apparently, the teacher did little to frame the lesson for the students, even though at the beginning of the case she noted several good reasons for doing the pattern-block activity. The main idea of the lesson—that one can arbitrarily choose any block to represent a whole and that the values of the other blocks depend on which block is designated the whole—was not explained directly. What could the teacher have said or asked to help students understand the lesson's purpose?

▪ *Fill-in-the-blank teaching*

In an effort to make the lesson easier for students, the teacher chose to do a directed lesson. The students seemed to follow along, answering her questions as she took them step by step through each example. Yet the issue raised at the end of the case may make us question the interactions between teacher and students. What did the teacher mean when she said students could only name or write the fractions if given a fill-in-the-blank question?

The students did fine as long as the teacher asked simple questions like, "How many reds make a yellow?" or "How many thirds is this?" Notice that the teacher was the one to say that the red represented halves and that the blue represented thirds. She "reminded" students that the yellow was equal to 1 whole and that, because 2 reds made 1 yellow, 2 was the denominator. Students evidently did not make that connection themselves, perhaps because they were not encouraged to process the information. When asked about the value of the green triangles, the students responded, "Two." This should not have been surprising, since the discourse had actually reinforced whole-number rather than fraction responses.

■ *Turn it into a problem-solving experience?*

What is the consequence of "making it simple" for the students? In this case, the teacher thought she was helping students by carefully guiding them through the steps. But what if the teacher had opened up the activity and made it a problem-solving experience? She could have asked students to work in groups to decide the names for each of the pieces, once it was established that the yellow was the whole. By working in groups, students could have helped each other deal with the terminology. A follow-up discussion would help students consolidate those ideas. By spoon-feeding students, the teacher may have actually reinforced the idea that they weren't capable of thinking for themselves.

■ *Too difficult for fifth graders?*

The task of filling out the chart is not as simple as it seems. For example, if the red block is 1, then the yellow block is 2, the blue block is $\frac{2}{3}$, and the green block is $\frac{1}{2}$. Provide pattern blocks for teachers to try this activity themselves. Is this activity appropriate or valuable for fifth graders? Can fifth graders do this as a problem-solving experience?

■ *Part-whole relationships?*

What are the limitations of pattern blocks for teaching fraction concepts? Because the divisions of the blocks are already made, students do not experience the partitioning process themselves. Also, the physical characteristics of the blocks make it easy for students to link the shape or color to the fraction rather than focusing on the part-whole relationship. The activity presented in this case is an excellent way to prevent this rote thinking, since the fractional name for each block changes with the designation of the whole.

Suggested Reading

Ball, D. L. 1991. "'What's All This Talk About Discourse?': Implementing the Professional Standards for Teaching Mathematics." *Arithmetic Teacher* 39(8): 14–48.

Donovan, B. F. 1990. "Cultural Power and the Defining of School Mathematics: A Case Study." In *Teaching and Learning Mathematics in the 1990s*, edited by T. J. Cooney and C. R. Hirsch, 166–173. Reston, VA: The National Council of Teachers of Mathematics.

Yackel, E., P. Cobb, T. Wood, G. Wheãtley, and G. Merkel. 1990. "The Importance of Social Interaction in Children's Construction of Mathematical Knowledge." In *Teaching and Learning Mathematics in the 1990s*, edited by T. J. Cooney and C. R. Hirsch, 12–21. Reston, VA: The National Council of Teachers of Mathematics.

What Next?

Where Do I Go from Here?

The author of this case raises questions that many teachers raise when they consider incorporating writing into their mathematics instruction. This case discussion allows teachers to analyze student writing and consider its value for further discussion.

Questions and Issues Raised by the Case

▪ *Diversity of student work*

It is interesting to note that while the instruction was the same for all students in the class, they exhibited many different ways of understanding—or misunderstanding—the learning experiences. Writing is a good way for teachers to see that teaching and learning is an interactive, unpredictable process. It is also a way for students and the teacher to communicate about what is and isn't understood. One mark of a good writing task is whether it provokes a range of solutions and solution methods.

▪ *Discussion follow-up*

Instead of seeing the diversity of student responses as a problem, the teacher could have viewed them as an opportunity for further learning. Asking students to present and discuss their drawings and writings, even if flawed, is one way to expand the learning experience. For this to be successful, students must understand that the purpose of the discussion is to ask questions, analyze, and learn from each others' mistakes and correct strategies, rather than to show off or destructively criticize.

▪ *What do students understand?*

Trying to figure out students' thinking from their writings and drawings is an interesting exercise. Have teachers examine Joe's drawings, for example. Why might he have drawn a huge square for the smaller decimal and a tiny square for the larger decimal? Did Jake's number line indicate that he understood? Why might Carolyn have said that thousandths are bigger than tenths, even though her drawing was accurate? Could it be that she was getting interference from her knowledge of whole numbers where thousands are greater than tens? What did Randy mean when he said that "0.292 means 2 and 92 thousand(th)s?" Was he viewing each of his candy bars as a tenth rather than a whole, so that 0.292 was 2 tenths (2 candy bars) and 92 thousandths (part of another candy bar)?

- ***Stimulate or stifle creativity?***

George has an inventive mind. How would you respond to his drawings? Of course, it is difficult to answer this question without knowing George, but teachers in the case discussion might share their own experiences with "Georges." One thing to consider is that creativity is sometimes stifled in mathematics class, whereas it should be encouraged. Perhaps all George needed was a little prodding to share his thoughts in writing, or perhaps he needed more practice and feedback to develop these skills. After all, the students in this class were relatively inexperienced writers, according to the case author.

Suggested Reading

Clarke, D. J., D. M. Clarke, and C. J. Lovitt. 1990. "Changes in Mathematics Teaching Call for Assessment Alternatives." In *Teaching and Learning Mathematics in the 1990s,* edited by T. J. Cooney and C. R. Hirsch, 118–129. Reston, VA: The National Council of Teachers of Mathematics.

Sullivan, P., and D. Clarke. 1991. "Catering to All Abilities Through 'Good' Questions." *Arithmetic Teacher* 39(2): 14–18.

Favorite-Food Circle Graph

The intention of the teacher in this case was to have students relate fractions, decimals, ratios, and percents, but the focus of the lesson was on finding a percent given a ratio. The lesson context was making a survey from a circle graph. While the context provided a base for reality, it also introduced some interference into the lesson.

Suggested Materials: Calculators.

Questions and Issues Raised by the Case

- ***Students' interpretation of percent***

Looking at the key on the circle graph, how did students interpret each section of the graph? It appears that they may have been thinking of the whole as 360 degrees and each of the 36 sections as $\frac{1}{10}$ of the whole or 10%. When you realize that students had to maintain and operate with two versions of the whole, one for the number of degrees in the circle and the other for the number of people in the survey, it is understandable why the confusion might have occurred.

- ***Teacher's intentions***

What was the purpose of this lesson? The teacher stated that the purpose of the unit was to help students realize that fractions, decimals, ratios, and percents were all ways to express parts of a whole. Was this a reasonable goal for sixth-grade students? Why might the teacher have decided to do this in an introductory unit on percent? What are the possible consequences if these ideas are not interrelated? Is it better to integrate the concepts later on rather than in a beginning unit on percent?

- ***Meaning of percent***

In the second lesson, the teacher talked about the relationship between percents and decimals, saying that since 100% is all the people in the survey then 1% is the same as one hundredth. She then had students find the decimal equivalent for each ratio in the survey to the nearest hundredth and helped them reason that 0.38 would be the same as 38%. Do you think most students could follow her reasoning? What might have led students to decide that 0.2 was 2%? Is there a better way to help students understand the relationship between fractions, decimals, ratios, and percents?

Is it possible that the students needed to develop some basic ideas of percent before this lesson? For example, did students realize that $\frac{22}{36}$ is more than $\frac{1}{2}$ and therefore more than 50%? Did they realize that if $\frac{2}{36}$ was 6%, logically $\frac{1}{36}$ should be about 3%? How did they justify (Or did they even question?) that $\frac{1}{36}$ was 3% and that $\frac{3}{36}$ is 8%? How might estimation and prediction be built into this lesson to make students focus on the broader notions of percent, rather than a calculator rule for changing a fraction to a decimal to a percent?

■ *Role of the calculator in concept development*

What part did the calculator play in this lesson? Did it help or hinder the students in learning about the meaning of percent? Was the point of this lesson to help students understand that $\frac{22}{36}$ can be expressed as 61% by plugging in the numbers into the calculator and getting 0.6111111? Would they have known the difference if the calculator had said 0.1111111? What was needed in this lesson so that students knew whether or not their answers make sense? Or does this kind of understanding develop through continued use of the calculator and additional experiences with percent? Did the calculator help students focus on the concept or simply give them a tool to find the answer to their problems?

Suggested Reading

Allinger, G. D., and J. N. Payne. 1986. "Estimation and Mental Arithmetic with Percent." In *Estimation and Mental Computation,* edited by H. L. Schoen and M. J. Zweng, 141–155. Reston, VA: The National Council of Teachers of Mathematics.

What's My Grade?

National assessments indicate that many of our students can apply the rules but do not have the underlying concepts for percent. In this case, the teacher recognized a common misconception that students have about percent and wondered how to alter her percent unit to help better prepare her students to avoid or work through misconceptions such as these.

Suggested Materials: Grid paper, calculators.

Questions and Issues Raised by the Case

- ***Percent as a ratio***

What is the misunderstanding? It is common for students to reason that missing two problems is not very many, therefore a C seems unreasonable. What Phillip didn't realize was that he also needed to pay attention to the number of problems on the whole test. It is a comparison or ratio of two numbers that gives us a percent, rather than the just the number of problems missed. What could have been said to Phillip that might have helped him see that his reasoning was flawed or incomplete?

- ***Introducing percent as "so many" out of one hundred***

The model this teacher used to introduce percent—coloring a given number of squares out of a grid of 100 squares—is a common one. Analyzing the model may raise some questions. For example, will students always think that they have to have 100 things in order to figure how what the percent is? Would they ever understand that 40 out of 100 is the same percentage as 20 out of 50? Does this model help students to see that missing 2 out of 8 problems isn't as close to 100% as missing 2 out of 80 problems or 2 out of 100 problems? Could this have any bearing on creating or avoiding misconceptions such as the one expressed by Phillip?

- ***Alternative models?***

One way to think about percent is in terms of fractions: One whole is 100%, one half is 50%, and one tenth is 10%. These benchmark percents can then be used to find other percents by using reasoning. Is there an advantage or disadvantage to relating fractions to percents in this way early in the conceptual development process? How would you do it?

Ideas such as the following might arise from a discussion of this issue. Start with the premise that if there are 10 problems on a test, then 10 correct problems represents 100%.

> *How many problems would be correct if you had 50%? (Remember that 50% is the same as a half.)*
> *How many problems would be correct if you had 10% (or one-tenth)?*
> *Could you get 25% on the test? Why or why not? (Since 50% is 5 problems, 25% would be half of that or 2.5 problems. Perhaps you could get partial credit on the problems.)*

What other models might be better? Could the square-grid model be altered so that students could figure out what percent 6 out of 8 problems would be? A dollar bill is often offered as an alternative to the square grid. What are the advantages and disadvantages of using this model? What about relating percentages to circle graphs?

■ *Enough time?*

Do you think the students in this case could develop a basic understanding of percent concepts in 4 or 5 days? What should the goals for understanding be? How do we know if students understand percent? What kinds of problems would they be able to do? What kinds of misconceptions would we hope to avoid? Why was it that this teacher's students still didn't understand that missing 2 out of 8 led to a lower score than 2 out of 40? Are students capable of understanding percent at this age?

Suggested Reading

Allinger, G. D., and J. N. Payne. 1986. "Estimation and Mental Arithmetic with Percent." In *Estimation and Mental Computation*, edited by H. L. Schoen and M. J. Zweng, 141–155. Reston, VA: The National Council of Teachers of Mathematics.

Connections

Percents, Proportions, and Grids

After having difficulty teaching students how to solve percent problems in the past, the teacher in this case encouraged students to solve percent problems by using a 10-by-10 grid and setting up proportions. In spite of the teacher's new approach, students were unable to solve the same types of problems on the following day. This case provides teachers with the opportunity to discuss the teaching of percents and the relative utility of using proportions, equations, or other methods to help students develop an understanding of percents.

Suggested Materials: Grid paper, calculators, transparency of grid.

Questions and Issues Raised by the Case

■ *Relevancy to students' lives*

How could the teacher have made these exercises more relevant for students? Would they have made more sense to students if they were in the context of a word problem?

Are all of these types of percent problems equally relevant or are some more common than others? For example, the first type of problem is relatively easy to link to real situations: "A $20 shirt is marked down 15%. What is the discount?" The second type of problem is when one is asked to find the percent discount for a $20 shirt that was marked down $3. The third type of problem is when one finds the original price of a shirt knowing that it was marked down $3, which was a 15% discount. Should all three types of percent problems be taught in sixth grade?

■ *Percent sense*

Does this method promote an intuitive sense of percent? Does it impede the development of intuitive approaches? What if students were asked first to do very simple percent problems and then to see if they could apply that method to more difficult examples? For example:

What is 50% of 20?
What % of 20 is 10?
10 is what % of 20?

In what way are all of these problems the same problem? Could students use reasoning and what they already know about benchmark fractions to build their sense of percent?

If 20 dots represent 100%: •
Find 50% of 20 dots. (If 100% is the whole thing, then 50% must be half or 10).
Find 10% of 20. (Divide the dots into tenths, so 10% is 2 dots.)
What percent would 5 dots be? (10 dots is 50%, so 5 dots is 25%.)
What percent would 4 dots be? (2 dots is 10%, so 4 dots is 20%.)
What percent would 3 dots be? (2 dots is 10%, so 3 dots is 5% more or 15%.)

The question is whether or not the students can transfer what they understand with simple percents to more complicated percents. Is it better to just have a rule to follow?

▪ *Proportions or equations?*

Which is the better method for representing percent problems, proportions or equations? What issues need to be considered with each approach? Would one of these methods be better for older students and one better for younger students? Which method ties best to our natural language? Note that some educators teach students to translate the language into an equation: "15% of 20 is what?" becomes "$15\% \times 20 = n$"; or "3 is 15 percent of what?" becomes "$3 = .15 \times n$." What are the best models for these approaches? Will the grid work with both?

▪ *Student inventions or rote rules?*

Is there a way the teacher could have turned this lesson into a discovery or problem-solving lesson? Is it possible for students invent their own approaches? What problem or situation might be posed for students to test out their own ideas for interpreting and solving percent problems?

Suggested Reading

Allinger, G. D., and J. N. Payne. 1986. "Estimation and Mental Arithmetic with Percent." In *Estimation and Mental Computation*, edited by H. L. Schoen and M. J. Zweng, 141–155. Reston, VA: The National Council of Teachers of Mathematics.

Janvier, C. 1990. "Contextualization and Mathematics for All." In *Teaching and Learning Mathematics in the 1990s*, edited by T. J. Cooney and C. R. Hirsch, 183–193. Reston, VA: The National Council of Teachers of Mathematics.

Six Hours Isn't One-Sixth of a Day

The teacher attempted to get students to understand the relationship between pictorial representations of fractions on a circle graph and the symbolic representation of fractions. When the teacher asked the class what fraction was represented by 6 pieces (of a circle divided into 24 parts), no one gave the correct answer. This case provides an opportunity to discuss the use of fraction kits and circle graphs for teaching the naming of fractions and the equivalency of fractions.

Suggested Materials: Fraction pieces.

Questions and Issues Raised by the Case

▪ *Transfer from the fraction kit*

Many teachers identify with the problem of helping students transfer what they learn in one situation to another. In this case, why didn't the students transfer what was learned with the fraction kit to the circle graph? One possibility is that students really didn't form deep conceptual understandings by using the fraction kit. Activities to establish an understanding of equivalent fractions using the fraction kit, for example, can be performed by simply matching 2 blue pieces to 1 red piece. Although students know and use the fraction names for the blue and red pieces, the conceptual underpinnings may be absent. They need to explore why $\frac{2}{4}$ is the same as $\frac{1}{2}$, not just say that it is so because "they fit."

Another reason that the fraction kit may not transfer well to the circle graph is that they are modeling different things. The fraction kit models fractions of a continuous whole unit, while the circle graph models fractions of a set of units. In other words, the circle graph shows what fraction of 24 hours you sleep or eat. By displaying the 24 hours in a circle graph, the distinction between finding a fraction of 24 hours, finding a fraction of the whole circle, and finding a fraction of a set is blended. It is not clear whether this would have helped or hindered students. Generally, however, finding a fraction of a set of things is considered to be conceptually more difficult that finding a fraction of one continuous thing.

There may be other reasons that the students did not transfer the idea of equivalence. For example, some fraction kits are limited to denominators of 2, 4, 8, and 16. Students may not have experienced twenty-fourths, sixths, or thirds in their fraction kits and had no mental tools for figuring out the equivalencies.

What are some things that students might do to develop a deeper understanding of equivalence? Students need lots of practice subdividing units into fractional parts to develop an understanding of how the parts relate to the whole and to each other. Doing the subdividing once with the fraction kit is not enough to develop this understanding. They should also be asked to show and explain why $\frac{6}{8}$ is equivalent to $\frac{3}{4}$ or $1\frac{1}{2}$ is equivalent to $\frac{3}{2}$.

Experiences are also needed with many different shapes and kinds of units. Students need to subdivide a square, a circle, a length of string, an orange, a jar of water, and so on. Measuring with a ruler or using measuring cups for cooking can give practical experiences with fractions.

▪ *Developmental concerns*

Are fourth graders ready for fraction equivalency? Are they capable of understanding part-whole relationships at this age? Although students are capable of forming informal understandings about fractions at this age, their work with fractions should focus on building an intuitive base connected to their own experiences. Learning how to draw, name, count, and estimate fractional parts is more appropriate than learning formal rules and computing. This particular activity did connect fractions to students' experiences informally, but it may have been a little difficult for fourth graders.

▪ *The rule for naming fractions*

Students, like Karen, come up with many logical ways for naming fractions. ("Six hours equals one-sixth.") Most of them have been taught rules for naming and writing fractions like: "Count the number of equal parts in the circle and write that number as the denominator. Count the number of parts shaded and write that number in the numerator." How might this rule have interfered with the students' ability to name fractional parts in the circle graph? Notice that the circle graph displayed both equal and unequal parts. There were 6 equal parts in the green part of the graph that Karen called $\frac{1}{6}$. Perhaps it made sense to a student to call this part $\frac{1}{6}$.

▪ *Visualizing fractions*

Why didn't the students visually recognize that the amount of time spent in school was $\frac{1}{4}$ of the circle? Would they have been able to recognize this if the circle were turned so that the $\frac{1}{4}$ section was in a more typical orientation? Should students have more experience naming fractions that are not in the typical orientation? Knowing that a part is about $\frac{1}{3}$ or about $\frac{1}{4}$ by visualizing is a valuable skill.

It is used, for example, to interpret the information displayed in graphs or to estimate measures. What kinds of experiences are needed to develop this skill?

Suggested Reading

Behr, M. J., T. R. Post, and I. Wachsmuth. 1986. "Estimation and Children's Concept of Rational Number Size." In *Estimation and Mental Computation*, edited by H. L. Schoen and M. J. Zweng, 103–111. Reston, VA: The National Council of Teachers of Mathematics.

Woodcock, G. E. 1986. "Estimating Fractions: A Picture is Worth a Thousand Words." In *Estimation and Mental Computation*, edited by H. L. Schoen and M. J. Zweng, 112–115. Reston, VA: The National Council of Teachers of Mathematics.

This Wasn't My Plan

Bubbles to Kickball

Part One

The well-intentioned beginning teacher in this case made some mistakes to which most experienced teachers will relate. However, teachers should not dismiss what they can learn from an analysis of the case, which offers an opportunity for teachers to clarify their own understandings of average and remainders, as well as a chance to confront gender issues.

Suggested Materials: Calculators, decimal place-value squares, base-10 blocks, string, tape measures.

Questions and Issues Raised by the Case

▪ *Student errors*

Why do students make errors such as getting an answer of 6.1 when dividing 25 by 4? Is this simply a careless error? Because the error was consistent among so many students in this case, some rationality was likely behind it. When students encounter a situation they don't know how to handle, they usually look for the most logical solution. In this case, they knew that the remainder was placed next to the quotient when decimals weren't involved, so it seemed reasonable to put the remainder in the same place when decimals were involved. They knew that with decimals you didn't need to write the *r*, so they just made a decimal point and wrote the remainder.

▪ *What does the remainder mean?*

What is the meaning of the remainder in this problem? Does it mean 1 divided into 4 parts? Or could it be 1 out of 4 things? Or are we asking what part 1 is of 4? Is it true, as Reggie said, that "one is 25 hundredths of 4"? Or should he have said, "One divided by 4 is 25 hundredths"?

It may help to think about what the problem means. The students were dividing a length of 25 centimeters into 4 equal parts. Each part was 6 centimeters long, with 1 centimeter left to divide equally 4 ways ($1 \div 4$). So, the length for each of the 4 parts was $6\frac{1}{4}$ or 6.25.

▪ *Concept of average*

Students in this lesson showed that they knew which algorithm to use to find the average. However, this may not mean they understood what the average meant. How could a teacher have helped them understand the process of finding an average? The students could have simply measured out string to match the length of each diameter. Then, all the lengths could have been combined and the string folded into fourths to find the average. The measure of the folded string should have approximately matched the average found by computation.

Another broader way of thinking about the meaning of average is to think of it as an equalizing process. If a little of the diameter from one bubble was given to a smaller bubble until all of the bubble diameters were equal, what would their diameters each have been? If the diameters were matched to string, the process could actually be modeled, although the accuracy of this method would be questionable.

Part Two

In this second part of the case, the teacher extended a discussion on averages to each individual's kicking averages in a kickball game. Though the teacher hoped that using and calculating averages in this context would aid student understanding, it seemed to have the opposite effect. This case provides the opportunity to discuss "subtle" distinctions between the problems given by the teacher in Part One of this case and those given in Part Two of the case. These subtle distinctions could have had a large impact on the students' ability to transfer what was learned in one problem to another.

Questions and Issues Raised by the Case

▪ *Gender issues*

Although students in the class all played kickball and the lesson did relate to their lives, as the beginning teacher intended, there are still some gender issues to consider. How many of the students were familiar with baseball statistics? It is not surprising that Rigo, a male student, connected to the example given by the teacher. Although the case doesn't say, one might guess that many of the female students were less likely to relate to the example. What problem situation would have had a better chance of relating to all students, regardless of their gender, ethnicity, or other characteristics?

▪ *Quotient less than one or greater than one?*

Might there have been a better real-life situation that the teacher could have used to help students understand the concepts? One dimension that the teacher did not plan for was the mathematical difference between the two examples. In the first lesson, students divided a larger number by a smaller number, so that the quotient was greater than 1. In the second lesson, a smaller number was divided by a larger number and the quotient was less than 1. In the second lesson, the students also had to add a decimal point and zeros before they could start dividing. This may have increased the confusion for students and made it more difficult for them to make sense of their answers.

▪ *Meaning of kickball averages versus bubble averages*

To students, kickball averages may have a different connotation than bubble averages. With bubble averages you are looking for the "middle" size of the bubbles, whereas with kickball averages you are trying to determine what 1 out of 4 (1 ÷ 4) hits would be. Although you can think of kickball averages as the "middle" number of hits you make per number of attempts, an average of .250 is more commonly thought of as a number on a scale from 0 to 1000, where 300 is considered to be a pretty good average.

Suggested Reading

Rowan, T. E., and N. D. Cetorelli. 1990. "An Eclectic Model for Teaching Elementary School Mathematics." In *Teaching and Learning Mathematics in the 1990s,* edited by T. J. Cooney and C. R. Hirsch, 62–68. Reston, VA: The National Council of Teachers of Mathematics.

The Beauty of Math

Various reform agenda encourage the presentation of mathematics in a way that enables students to appreciate the beauty of the discipline. In this case, the teacher attempted to facilitate an appreciation of the beauty of mathematics by having students explore problems that incorporated mathematical patterns and elegance. Students solved the problem in various ways, but many seemed to miss the beauty by getting lost in the mechanics of the computation.

Questions and Issues Raised by the Case

▪ *What did the teacher want?*

In this case, the teacher presented a problem by asking the students to simplify the following expression.

$$\frac{1}{1\cdot 2}+\frac{1}{2\cdot 3}+\frac{1}{3\cdot 4}+\frac{1}{4\cdot 5}+\frac{1}{5\cdot 6}$$

For several math periods prior to the lesson in this case, the students had explored expressions that were directly related to this problem. In this lesson, the teacher wanted students to substitute expressions from the previous explorations for the terms in this new problem. Even though the teacher had devoted considerable time to these explorations, students still did not see the relationship and used other less elegant methods to solve the problem. Why did this happen?

▪ *Are the expressions equivalent?*

Did students understand that the expressions they were expected to substitute were equivalent to the expressions in the new problem? The following are the kinds of substitutions the teacher had in mind for each term in the problem above.

First term in the new problem:	Substitution:
$\frac{1}{1\cdot 2}$	$\frac{1}{1}-\frac{1}{2}$
Second term in the new problem:	Substitution:
$\frac{1}{2\cdot 3}$	$\frac{1}{2}-\frac{1}{3}$

To find out if the first term of the original problem is equivalent to its substitute, you can simplify $\frac{1}{1 \cdot 2}$ to get $\frac{1}{2}$. Then, you can solve $\frac{1}{1} - \frac{1}{2}$ mentally: $1 - \frac{1}{2}$ is also $\frac{1}{2}$. Therefore both expressions equal $\frac{1}{2}$.

To find out if the second term is equivalent to its substitution, you simplify $\frac{1}{2 \cdot 3}$ to get $\frac{1}{6}$. To simplify $\frac{1}{2} - \frac{1}{3}$, you rewrite the fractions with common denominators: $\frac{3}{6} - \frac{2}{6}$ is $\frac{1}{6}$. These expressions are also equal.

Is it too much to expect seventh- and eighth-grade pre-algebra students to understand this mathematics? Is it appropriate for students who are not considered gifted and talented?

■ *Why are the expressions equal?*

What is it about the mathematics that makes these substitutions work? Consider the following expressions.

$$\frac{1}{1} - \frac{1}{2}$$

$$\frac{1}{2} - \frac{1}{3}$$

$$\frac{1}{3} - \frac{1}{4}$$

You might ask students to think about how to solve these problems. Notice that to find common denominators for each pair of fractions you can multiply the original denominators together. So, for the following fractions 6 is the common denominator.

$$\frac{1}{2} - \frac{1}{3}$$

To change both fractions into new fractions with the common denominator of 6, the numerator of the first fraction will be multiplied by 3 and the numerator of the second fraction with be multiplied by 2. The new fractions are:

$$\frac{3}{6} - \frac{2}{6} = \frac{1}{6}$$

Note the new numerators are the original denominators of the original pair of fractions, except their order is reversed. This happens because the original numerators are 1.

Is it important for students to understand why this works? How would a teacher help students figure this out?

■ *Why is substitution helpful in this problem?*

Often substitution of equivalent expressions is performed in mathematical investigations. Substitutions can serve to place all terms in common form (i.e., so that apples are not subtracted from oranges). Substitution can also serve to simplify an expression. An example of the first type of substitution is typically found when borrowing is performed during a subtraction problem. For example, to subtract 83 from 90 one borrows a set of 10 ones from the 8, and the subtraction then becomes 3 from 10.

The teacher demonstrated substitution of the second type when replacing the expression—

$$\frac{1}{1 \cdot 2} + \frac{1}{2 \cdot 3} + \frac{1}{3 \cdot 4} + \frac{1}{4 \cdot 5} + \frac{1}{5 \cdot 6}$$

—with the following equivalent expression.

$$\left(\frac{1}{1} - \frac{1}{2}\right) + \left(\frac{1}{2} - \frac{1}{3}\right) + \left(\frac{1}{3} - \frac{1}{4}\right) + \left(\frac{1}{4} - \frac{1}{5}\right) + \left(\frac{1}{5} - \frac{1}{6}\right)$$

This expression without the parenthesis is as follows.

$$\frac{1}{1} - \frac{1}{2} + \frac{1}{2} - \frac{1}{3} + \frac{1}{3} - \frac{1}{4} + \frac{1}{4} - \frac{1}{5} + \frac{1}{5} - \frac{1}{6}$$

All of the terms in this expression, except for $\frac{1}{1}$ and the $-\frac{1}{6}$, are being both added and subtracted and thus "cancel" each other out. All that is left is $\frac{1}{1}$ and $-\frac{1}{6}$.

$$\frac{1}{1} \underbrace{- \frac{1}{2} + \frac{1}{2}}_{+0} \underbrace{- \frac{1}{3} + \frac{1}{3}}_{+0} \underbrace{- \frac{1}{4} + \frac{1}{4}}_{+0} \underbrace{- \frac{1}{5} + \frac{1}{5}}_{+0} - \frac{1}{6}$$

$$\frac{1}{1} \quad +0 \quad +0 \quad +0 \quad +0 \quad - \frac{1}{6}$$

$$\frac{1}{1} - \frac{1}{6}$$

This substitution simplifies the problem into $1 - \frac{1}{6}$ or $\frac{5}{6}$.

■ *Cancellation of inverses*

The teacher demonstrated how to "cancel out" a number by adding its opposite or inverse, as is illustrated in the following expression.

$$\left(\frac{1}{1} - \frac{1}{2}\right) + \left(\frac{1}{2} - \frac{1}{3}\right) + \left(\frac{1}{3} - \frac{1}{4}\right) + \left(\frac{1}{4} - \frac{1}{5}\right) + \left(\frac{1}{5} - \frac{1}{6}\right)$$

The negative one-half inside the first parenthesis and the positive one-half inside the next parenthesis cancel one another out since all the terms are being added together.

Why is the process of canceling opposites beneficial? Is it possible that teaching these students to cancel out opposites is problematic? Without careful observation, students can receive the impression that numbers are magically dropping out of the problem. Also, in this problem the canceling of opposites necessitates canceling numbers from different sets of parentheses. This can be problematic to students if the only way they know is first to simplify within each parenthesis and then add the results.

Students' solutions or teacher's solution?

While students were allowed to choose the method that they would use to solve the problem, most of them chose to find common denominators and perform the subtraction. Should students be expected to come up with the teacher's solution, or is it appropriate for them to have their own solutions, even though the solution might be considered less beautiful in the eyes of a mathematician? Why is it that students often fall back on algorithms as a reliable tool without looking for the beauty of the math?

What about the group that got 70 as an answer? What were they thinking about? What could be done to address their misconceptions? Their solution technique involved adding denominators as if they were whole numbers. Why would students do this? One possibility is that they were confusing multiplication and division by inverting and multiplying.

Consecutive and sequential numbers

When describing unit fractions, the teacher used the words *consecutive* and *sequential numbers*. What are consecutive and sequential numbers? Sequential numbers are numbers that follow one after another in order. Consecutive numbers also follow one after the other in order but without gaps. Some teachers might debate whether or not these concepts and their distinctions are worth spending time on, and yet, given the increased emphasis on recognizing patterns in the current math reform, these ideas take on increased importance. In addition, the mathematics these students will learn in the future—particularly calculus, probability, and statistics—center on these concepts.

Beauty and connectedness

What did the teacher mean by the words *connectedness* and *beauty* in mathematics? He mentioned that the problem he assigned included commutative and associative properties, the use of different mathematical operations, patterns, and inverse operations. Also, mathematicians sometimes refer to proofs and expressions as beautiful when complex problems are solved with elegance and simplicity. Given the teacher's description of the presentation, in what ways, if at all, might the students have experienced these problems as connected or beautiful? How can this message be relayed to students so that they appreciate the beauty? How is beauty relevant to math?

Suggested Reading

Steffe, L. P. 1990. "Adaptive Mathematics Teaching." In *Teaching and Learning Mathematics in the 1990s,* edited by T. J. Cooney and C. R. Hirsch, 41–51. Reston, VA: The National Council of Teachers of Mathematics.

The Ratio of Girls to Boys

The teacher introduced the concept of ratio by comparing the number of girls to the number of boys in the classroom. Some of the students claimed that because the ratio was greater than 1, and because their classroom was 1 classroom, the teacher's ratio was incorrect. The students were confused about the similarities and differences between fractions and ratios.

Suggested Materials: Counters.

Questions and Issues Raised by the Case

▪ *Four meanings of ratio*

The teacher stated that the fraction $\frac{1}{2}$ was a ratio that meant 1 divided by 2 or 1 *out of* 2. Are there additional meanings for ratio? How is a ratio different from a fraction? This case provides an opportunity for discussion participants to think about the different ways ratios are expressed. Some of those ways are listed below.

1 *out of* 2
1 *to* 2
1 *for every* 2
1 *divided by* 2

Commonly, the terms *out of, to,* and *for every* are interchanged. Conceptually, however, the meanings of these terms can be distinguished. Distinctions between each of these terms are discussed below; visual representations of these distinctions are also provided.

▪ *Part of a whole:* **1 *out of* 2**

The teacher pointed out that fractions, which are ratios, can describe a comparison between a part and a whole. For example, the ratio 1:2 could mean that 1 out of 2 cars are red, or it could mean that 50 out of 100 cars are red. In the case, we could say that 17 out of 32 students were girls or that 15 out of 32 students were boys. These comparisons show parts of a whole. In other words, 17 (a part) out of 32 (the whole class) were girls and 15 (a part) out of 32 (the whole class) were boys.

How might this kind of ratio be represented? Consider the ratio of 1 square shaded out of 2. One possible way to represent this visually is:

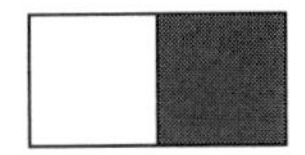

Comparison of two parts: **1** *to* **2**

Another way to use a ratio is to compare two parts. Consider the ratio of 1 apple to 2 oranges. It could mean that there is 1 apple and 2 oranges, 5 apples and 10 oranges, or 20 apples and 40 oranges.

In the case, the ratio of girls to boys was 17 to 15. This ratio was a comparison of 2 parts, 17:15. Unlike the *out of* ratio examples, the group of boys and the group of girls were 2 nonoverlapping sets.

In the case, we can infer that the whole class had 32 students since there were 17 girls and 15 boys. It is not always true that a ratio provides information about an entire set of things. Suppose that hair color, rather than gender, was expressed as a ratio. Consider, for example, "the ratio of blondes to brunettes in the classroom is 12 to 13." In this situation, we couldn't make any assumptions about the number of people in the whole class, since we wouldn't know how many students were neither blonde nor brunette.

How might this kind of ratio be represented? Consider the ratio of 1 white square to 2 shaded squares. One possible way to represent this visually is:

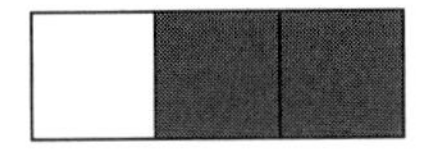

Proportion: **1** *for every* **2**

The meaning of 1 *for every* 2 is very similar to the ratio 1 *to* 2. However, common usage of these terms may imply a somewhat distinct representation for each. The ratio 1 *for every* 2 could suggest that there are sets of things in the given ratio.

One possible visual representation of 1 *for every* 2 or 1 white square for every 2 shaded squares, is:

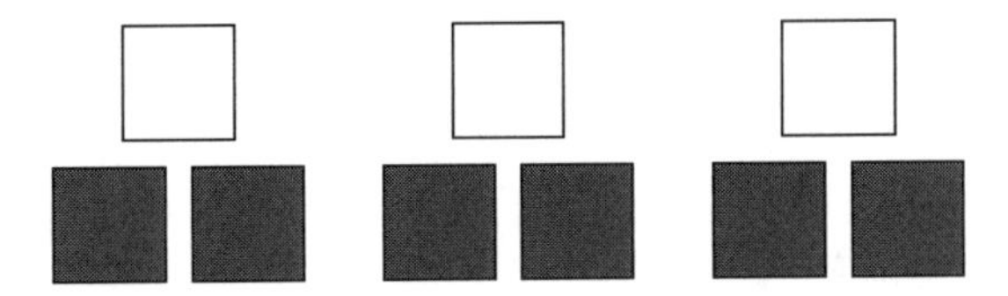

■ *Division:* 1 *divided by* 2

If you divide a whole into parts, you can describe the relationship between a whole and a part by a ratio that is also called a fraction. The teacher suggested that $\frac{1}{2}$ is 1 divided by 2. This could mean 1 whole square cut into two parts.

One possible visual representation of 1 *divided by* 2, or 1 square divided by 2, is:

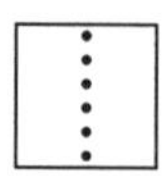

■ *Differences between ratios and fractions?*

Mathematicians may legitimately claim that there is no difference between ratios and fractions. However, fractions as they are commonly used in textbooks and classrooms are thought to express a relationship of parts to a whole. There are two forms of ratio that are fractions: 1 *out of* 2 and 1 *divided by* 2. Rarely, if ever, do the fractions represented by textbooks or teachers include examples of fractions which relate one part *to* another part (i.e., the ratio relationship usually expressed with the word *to*). Usually fractions indicate part-whole relationships.

■ *Simplifying and comparing ratios*

Ratios, expressed in lowest terms or simplest form, are usually the easiest to understand and compare. For example, it is easier to use and visualize 3:4 as a ratio than 27:36 or 42:56, even though these ratios could represent the same comparisons.

Another way that ratios can be more readily compared is by converting them into decimals or percents. Since $\frac{17}{32} = .53$ and $\frac{15}{32} = .47$, we could say that 53% of the whole class is made up of girls and 47% of the whole class is made up of boys.

■ *Using the example of the number of students in the class*

The teacher used an example of the ratio of girls to boys in the class. What is the impact of using the students themselves as the example? What are the benefits? Are there any drawbacks? Use of students in the problem could be a drawback if students forget to count themselves as members of the group. On the other hand, it could be positive in that students are comparing things that are in their immediate world.

Suggested Reading

Cramer, K., and T. Post. 1993. "Making Connections: A Case for Proportionality." *Arithmetic Teacher* 40(6): 342–346.

Appendix A

Case Discussion Process Assessment

Rating Scale:	1	2	3	4	5
	Strongly Disagree				*Strongly Agree*

1. ______ Members of the group were considerate of one another's feelings and opinions, even when disagreeing.
2. ______ The discussion was not dominated by anyone.
3. ______ Members built on and contributed to one another's ideas.
4. ______ Different points of view and alternative solutions were given respectful consideration, even when they might be considered by some to be "unpopular" or "not politically correct."
5. ______ The discussion had depth.
6. ______ There was time to reflect and collect one's thoughts.
7. ______ The discussion had satisfactory closure, even though hard and fast solutions or decisions may not have emerged.

Appendix B

Facilitator Feedback

Attentive Listening
How did the facilitator demonstrate attentive listening skills?

Balanced Participation
What specific techniques were used to balance participation among group members?

Safe Environment
What was done to promote an environment where contributions were valued and feelings respected? How was tension or discomfort handled?

Build on One Another's Ideas
What specifically did the facilitator do to help participants build on one another's ideas? Were participants asked to provide examples, clarifications, and reasons for statements? Did the facilitator use the issues generated by the group to guide the discussion?

Issues Addressed
To what degree were specific teaching, learning, *and* mathematical issues addressed? Did the facilitator have enough familiarity with the ideas to assist others in drawing them out?

Alternative Perspectives
How were alternative perspectives and solution strategies elicited for group consideration? Did the facilitator stop with the "right answer" or push for alternatives, even if unpopular?

Nonjudgmental Communication
What did the facilitator do to guide the discussion without being judgmental? How well did the facilitator refrain from judging participants' ideas or from promoting his or her own viewpoint?

Nonverbal Communication
Was the facilitator relaxed and poised? Was there a strong presence without signaling superiority? Were there distracting mannerisms that the facilitator should know about?

Appendix C

Advisory Board Members
Mathematics Case Methods Project

Phil Daro, Executive Director, California Mathematics Project, University of California Office of the President; Director for Assessment Development, New Standards Project, University of California Office of the President.

Lise Dworkin, Director, San Francisco Mathematics Collaborative.

Donna Goldenstein, Elementary Teacher, Hayward Unified School District.

Babette Jackson, Principal, Hayward Unified School District.

Carol Langbort, Chair, Department of Elementary Education, San Francisco State University.

Maisha Moses, Site Developer, The Algebra Project.

Judy Mumme, Director, Math Renaissance.

Sharon Ross, Professor, Department of Mathematics, California State University, Chico.

Jay Rowley, Principal, San Ramon Valley Unified School District.

Lee Shulman, Charles E. Ducommun Professor of Education, Stanford University.

Judy Shulman, Director, Institute for Case Development, Far West Laboratory.

Hardy Turrentine, Intermediate Teacher, Hayward Unified School District.

References

Allinger, G. D., and J. N. Payne. 1986. "Estimation and Mental Arithmetic with Percent." In *Estimation and Mental Computation*, edited by H. L. Schoen and M. J. Zweng, 141–155. Reston, VA: The National Council of Teachers of Mathematics.

Azzolino, A. 1990. "Writing as a Tool for Teaching Mathematics: The Silent Revolution." In *Teaching and Learning Mathematics in the 1990s*, edited by T. J. Cooney and C. R. Hirsch, 92–100. Reston, VA: The National Council of Teachers of Mathematics.

Ball, D. L. 1991. "'What's All This Talk About Discourse?': Implementing the Professional Standards for Teaching Mathematics." *Arithmetic Teacher* 39(8): 14–48.

Baroody, A. J. 1989. "Manipulatives Don't Come with Guarantees." *Arithmetic Teacher* 37(2): 4–5.

Behr, M. J., T. R. Post, and I. Wachsmuth. 1986. "Estimation and Children's Concept of Rational Number Size." In *Estimation and Mental Computation*, edited by H. L. Schoen and M. J. Zweng, 103–111. Reston, VA: The National Council of Teachers of Mathematics.

Beyer, B. 1983. "Common Sense About Teaching Skills." *Educational Leadership* 41(3): 44–49.

Borasi, R. 1990. "The Invisible Hand Operating in Mathematics Instruction: Students' Conceptions and Expectations." In *Teaching and Learning Mathematics in the 1990s*, edited by T. J. Cooney and C. R. Hirsch, 174–182. Reston, VA: The National Council of Teachers of Mathematics.

Carter, H. L. 1986. "Linking Estimation to Psychological Variables in the Early Years." In *Estimation and Mental Computation*, edited by H. L. Schoen and M. J. Zweng, 74–81. Reston, VA: The National Council of Teachers of Mathematics.

Clarke, D. J., D. M. Clarke, and C. J. Lovitt. 1990. "Changes in Mathematics Teaching Call for Assessment Alternatives." In *Teaching and Learning Mathematics in the 1990s*, edited by T. J. Cooney and C. R. Hirsch, 118–129. Reston, VA: The National Council of Teachers of Mathematics.

Cramer, K., and T. Post. 1993. "Making Connections: A Case for Proportionality." *Arithmetic Teacher* 40(6): 342–346.

Cuevas, G. 1990. "Increasing the Achievement and Participation of Language Minority Students in Mathematics Education." In *Teaching and Learning Mathematics in the 1990s*, edited by T. J. Cooney and C. R. Hirsch, 159–165. Reston, VA: The National Council of Teachers of Mathematics.

Curcio, F. R. 1990. "Mathematics as Communication: Using a Language-Experience Approach in the Elementary Grades." In *Teaching and Learning*

Mathematics in the 1990s, edited by T. J. Cooney and C. R. Hirsch, 69–75. Reston, VA: The National Council of Teachers of Mathematics.

Damarin, S. K. 1990. "Teaching Mathematics: A Feminist Perspective." In *Teaching and Learning Mathematics in the 1990s,* edited by T. J. Cooney and C. R. Hirsch, 144–151. Reston, VA: The National Council of Teachers of Mathematics.

Donovan, B. F. 1990. "Cultural Power and the Defining of School Mathematics: A Case Study." In *Teaching and Learning Mathematics in the 1990s,* edited by T. J. Cooney and C. R. Hirsch, 166–173. Reston, VA: The National Council of Teachers of Mathematics.

Driscoll, M., and B. Lord. 1990. "Professionals in a Changing Profession." In *Teaching and Learning Mathematics in the 1990s,* edited by T. J. Cooney and C. R. Hirsch, 237–245. Reston, VA: The National Council of Teachers of Mathematics.

Hiebert, J. 1984. "Children's Mathematical Learning: The Struggle to Link Form and Understanding." *The Elementary School Journal* 84(5): 497–513.

Hiebert, J. 1990. "The Role of Routine Procedures in the Development of Mathematical Competence." In *Teaching and Learning Mathematics in the 1990s,* edited by T. J. Cooney and C. R. Hirsch, 31–40. Reston, VA: The National Council of Teachers of Mathematics.

Hiebert, J. 1992. "Mathematical, Cognitive, and Instructional Analyses of Decimal Fractions." In *Analysis of Arithmetic for Mathematics Teaching,* edited by G. Leinhardt, R. Putnam, and R. A. Hattrup, 283–322. Hillsdale, NJ: Lawrence Erlbaum Associates.

Holmes, E. E. 1990. "Motivation: An Essential Component of Mathematics Instruction." In *Teaching and Learning Mathematics in the 1990s,* edited by T. J. Cooney and C. R. Hirsch, 101–107. Reston, VA: The National Council of Teachers of Mathematics.

Janvier, C. 1990. "Contextualization and Mathematics for All." In *Teaching and Learning Mathematics in the 1990s,* edited by T. J. Cooney and C. R. Hirsch, 183–193. Reston, VA: The National Council of Teachers of Mathematics.

Kamii, C. 1990. "Constructivism and Beginning Arithmetic, K–12." In *Teaching and Learning Mathematics in the 1990s,* edited by T. J. Cooney and C. R. Hirsch, 22–30. Reston, VA: The National Council of Teachers of Mathematics.

Long, M. J., and M. Ben-Hur. 1991. "Informing Learning Through the Clinical Interview." *Arithmetic Teacher* 38(6): 44–46.

Miller, L. D. 1993. "Making the Connection with Language." *Arithmetic Teacher* 40(6): 311–316.

National Council of Teachers of Mathematics. 1989. *Curriculum and Evaluation Standards for School Mathematics.* Reston, VA: The National Council of Teachers of Mathematics.

National Council of Teachers of Mathematics. 1991. *Professional Standards for Teaching Mathematics.* Reston, VA: The National Council of Teachers of Mathematics.

Passmore, J. 1967. "On Teaching to Be Critical." In *The Concept of Education*, edited by R. S. Peters. London: Routledge & Kegan Paul.

Rowan, T. E., and N. D. Cetorelli. 1990. "An Eclectic Model for Teaching Elementary School Mathematics." In *Teaching and Learning Mathematics in the 1990s*, edited by T. J. Cooney and C. R. Hirsch, 62–68. Reston, VA: The National Council of Teachers of Mathematics.

Schielack, J. F. 1991. "Reaching Young Pupils with Technology." *Arithmetic Teacher* 38(6): 51–55.

Secada, W. G. 1990. "The Challenges of a Changing World for Mathematics Education." In *Teaching and Learning Mathematics in the 1990s*, edited by T. J. Cooney and C. R. Hirsch, 135–143. Reston, VA: The National Council of Teachers of Mathematics.

Spiro, R. J., R. L. Coulson, P. J. Feltovich, and D. K. Anderson. 1988. "Cognitive Flexibility Theory: Advanced Knowledge Acquisition in Ill-Structured Domains." In *Tenth Annual Conference of the Cognitive Science Society*, 375–383. Hillsdale, NJ: Erlbaum.

Steen, L. A. 1990. "Mathematics for All Americans." In *Teaching and Learning Mathematics in the 1990s*, edited by T. J. Cooney and C. R. Hirsch, 130–134. Reston, VA: The National Council of Teachers of Mathematics.

Steffe, L. P. 1990. "Adaptive Mathematics Teaching." In *Teaching and Learning Mathematics in the 1990s*, edited by T. J. Cooney and C. R. Hirsch, 41–51. Reston, VA: The National Council of Teachers of Mathematics.

Stigler, J. W. 1988. "Research into Practice: The Use of Verbal Explanation in Japanese and American Classrooms." *Arithmetic Teacher* 36(2): 27–29.

Stigler, J. W., and H. W. Stevenson. 1991. "How Asian Teachers Polish Each Lesson to Perfection." *American Educator* (15)1: 12–47.

Sullivan, P., and D. Clarke. 1991. "Catering to All Abilities Through 'Good' Questions." *Arithmetic Teacher* 39(2): 14–18.

Webb, N., and D. Briars. 1990. "Assessment in Mathematics Classrooms, K–8." In *Teaching and Learning Mathematics in the 1990s*, edited by T. J. Cooney and C. R. Hirsch, 108–117. Reston, VA: The National Council of Teachers of Mathematics.

Wilde, S. 1991. "Learning to Write About Mathematics." *Arithmetic Teacher* 38(6): 38–43.

Witherspoon, M. 1993. "Fractions: In Search of Meaning." *Arithmetic Teacher* 40(8): 482–485.

Woodcock, G. E. 1986. "Estimating Fractions: A Picture is Worth a Thousand Words." In *Estimation and Mental Computation*, edited by H. L. Schoen and M. J. Zweng, 112–115. Reston, VA: The National Council of Teachers of Mathematics.

Yackel, E., P. Cobb, T. Wood, G. Wheatley, and G. Merkel. 1990. "The Importance of Social Interaction in Children's Construction of Mathematical Knowledge." In *Teaching and Learning Mathematics in the 1990s*, edited by T. J. Cooney and C. R. Hirsch, 12–21. Reston, VA: The National Council of Teachers of Mathematics.

For additional information about facilitator preparation and case-writing seminars, or to be placed on a mailing list for future publications of casebooks, contact:

Carne Barnett
Mathematics Case Methods Project
Far West Laboratory
730 Harrison Street
San Francisco, CA 94107-1242
Fax: 415-565-3012